GLUTAMATE-RELATED BIOMARKERS IN DRUG DEVELOPMENT FOR DISORDERS OF THE NERVOUS SYSTEM

WORKSHOP SUMMARY

Diana E. Pankevich, Miriam Davis, and Bruce M. Altevogt
Rapporteurs

Forum on Neuroscience and Nervous System Disorders

Board on Health Sciences Policy

INSTITUTE OF MEDICINE
OF THE NATIONAL ACADEMIES

THE NATIONAL ACADEMIES PRESS
Washington, D.C.
www.nap.edu

THE NATIONAL ACADEMIES PRESS • 500 Fifth Street, N.W. • Washington, DC 20001

NOTICE: The project that is the subject of this report was approved by the Governing Board of the National Research Council, whose members are drawn from the councils of the National Academy of Sciences, the National Academy of Engineering, and the Institute of Medicine.

This project was supported by contracts between the National Academy of Sciences and the Alzheimer's Association; AstraZeneca Pharmaceuticals, Inc.; CeNeRx Biopharma; the Department of Health and Human Services' National Institutes of Health (NIH, Contract No. N01-OD-4-213) through the National Institute on Aging, National Institute on Alcohol Abuse and Alcoholism, National Institute on Drug Abuse, National Eye Institute, NIH Blueprint for Neuroscience Research, National Institute of Mental Health, and National Institute of Neurological Disorders and Stroke; Eli Lilly and Company; Foundation for the National Institutes of Health; GE Healthcare, Inc.; GlaxoSmithKline, Inc.; Johnson & Johnson Pharmaceutical Research and Development, LLC; Lundbeck Research USA; Merck Research Laboratories; The Michael J. Fox Foundation for Parkinson's Research; the National Multiple Sclerosis Society; the National Science Foundation (Contract No. OIA-0753701); Pfizer Inc.; and the Society for Neuroscience. The views presented in this publication are those of the editors and attributing authors and do not necessarily reflect the view of the organizations or agencies that provided support for this project.

International Standard Book Number-13:   978-0-309-21221-2
International Standard Book Number-10:   0-309-21221-9

Additional copies of this report are available from The National Academies Press, 500 Fifth Street, N.W., Lockbox 285, Washington, DC 20055; (800) 624-6242 or (202) 334-3313 (in the Washington metropolitan area); Internet, http://www.nap.edu.

For more information about the Institute of Medicine, visit the IOM home page at: www.iom.edu.

Printed in the United States of America

The serpent has been a symbol of long life, healing, and knowledge among almost all cultures and religions since the beginning of recorded history. The serpent adopted as a logotype by the Institute of Medicine is a relief carving from ancient Greece, now held by the Staatliche Museen in Berlin.

Cover courtesy of: Dr. Ehud Isacoff, University of California–Berkeley.

Suggested citation: IOM (Institute of Medicine). 2011. *Glutamate-Related Biomarkers in Drug Development for Disorders of the Nervous System: A Workshop Summary.* Washington, DC: The National Academies Press.

"Knowing is not enough; we must apply.
Willing is not enough; we must do."
—Goethe

INSTITUTE OF MEDICINE
OF THE NATIONAL ACADEMIES

Advising the Nation. Improving Health.

THE NATIONAL ACADEMIES
Advisers to the Nation on Science, Engineering, and Medicine

The **National Academy of Sciences** is a private, nonprofit, self-perpetuating society of distinguished scholars engaged in scientific and engineering research, dedicated to the furtherance of science and technology and to their use for the general welfare. Upon the authority of the charter granted to it by the Congress in 1863, the Academy has a mandate that requires it to advise the federal government on scientific and technical matters. Dr. Ralph J. Cicerone is president of the National Academy of Sciences.

The **National Academy of Engineering** was established in 1964, under the charter of the National Academy of Sciences, as a parallel organization of outstanding engineers. It is autonomous in its administration and in the selection of its members, sharing with the National Academy of Sciences the responsibility for advising the federal government. The National Academy of Engineering also sponsors engineering programs aimed at meeting national needs, encourages education and research, and recognizes the superior achievements of engineers. Dr. Charles M. Vest is president of the National Academy of Engineering.

The **Institute of Medicine** was established in 1970 by the National Academy of Sciences to secure the services of eminent members of appropriate professions in the examination of policy matters pertaining to the health of the public. The Institute acts under the responsibility given to the National Academy of Sciences by its congressional charter to be an adviser to the federal government and, upon its own initiative, to identify issues of medical care, research, and education. Dr. Harvey V. Fineberg is president of the Institute of Medicine.

The **National Research Council** was organized by the National Academy of Sciences in 1916 to associate the broad community of science and technology with the Academy's purposes of furthering knowledge and advising the federal government. Functioning in accordance with general policies determined by the Academy, the Council has become the principal operating agency of both the National Academy of Sciences and the National Academy of Engineering in providing services to the government, the public, and the scientific and engineering communities. The Council is administered jointly by both Academies and the Institute of Medicine. Dr. Ralph J. Cicerone and Dr. Charles M. Vest are chair and vice chair, respectively, of the National Research Council.

www.national-academies.org

GLUTAMATE-RELATED BIOMARKERS IN DRUG DEVELOPMENT FOR DISORDERS OF THE NERVOUS SYSTEM PLANNING COMMITTEE[*]

DANIEL JAVITT (*Cochair*), New York University
CHI-MING LEE (*Cochair*), AstraZeneca Pharmaceuticals
HUDA AKIL, University of Michigan
MARK BEAR, Massachusetts Institute of Technology
JOHN DUNLOP, Pfizer
RICHARD FRANK, GE Healthcare, Inc.
WALTER KOROSHETZ, National Institute of Neurological Disorders and Stroke
MENELAS PANGALOS, AstraZeneca Pharmaceuticals
WILLIAM POTTER, Foundation for NIH Neuroscience Biomarker Steering Committee
RAE SILVER, Columbia University
NORA VOLKOW, National Institute on Drug Abuse
STEVIN ZORN, Lundbeck Research
STEPHEN ZUKIN, AstraZeneca Pharmaceuticals

Study Staff

BRUCE M. ALTEVOGT, Project Director, Institute of Medicine
SARAH L. HANSON, Associate Program Officer (until June 2010)
SARA SHNIDER, Christine Mirzayan Science and Technology Policy Graduate Fellow (until May 2010)
LORA K. TAYLOR, Senior Project Assistant, Institute of Medicine

---

[*] Institute of Medicine planning committees are solely responsible for organizing the workshop, identifying topics, and choosing speakers. The responsibility for the published workshop summary rests with the workshop rapporteurs and the institution.

v

INSTITUTE OF MEDICINE FORUM ON NEUROSCIENCE AND NERVOUS SYSTEM DISORDERS*

---

* Institute of Medicine forums and roundtables do not issue, review, or approve individual documents. The responsibility for the published workshop summary rests with the workshop rapporteurs and the institution.

WILLIAM THIES, Alzheimer's Association
NORA VOLKOW, National Institute on Drug Abuse
KENNETH WARREN, National Institute on Alcohol Abuse and
  Alcoholism
FRANK YOCCA, AstraZeneca Pharmaceuticals
STEVIN H. ZORN, Lundbeck USA
CHARLES ZORUMSKI, Washington University School of Medicine

IOM Staff

BRUCE M. ALTEVOGT, Forum Director
SARAH L. HANSON, Associate Program Officer (until June 2010)
DIANA E. PANKEVICH, Associate Program Officer (since October
  2010)
LORA K. TAYLOR, Senior Project Assistant
ANDREW POPE, Director, Board on Health Sciences Policy

Reviewers

This report has been reviewed in draft form by individuals chosen for their diverse perspectives and technical expertise, in accordance with procedures approved by the National Research Council's Report Review Committee. The purpose of this independent review is to provide candid and critical comments that will assist the institution in making its published report as sound as possible and to ensure that the report meets institutional standards for objectivity, evidence, and responsiveness to the study charge. The review comments and draft manuscript remain confidential to protect the integrity of the process. We wish to thank the following individuals for their review of this report:

Mark Geyer, University of California–San Diego
John Krystal, Yale School of Medicine
Herb Meltzer, Vanderbilt University School of Medicine
Kalpana Merchant, Eli Lilly & Company

Although the reviewers listed above have provided many constructive comments and suggestions, they did not see the final draft of the report before its release. The review of this report was overseen by **Dr. Joseph T. Coyle,** Harvard Medical School. Appointed by the Institute of Medicine, he was responsible for making certain that an independent examination of this report was carried out in accordance with institutional procedures and that all review comments were carefully considered. Responsibility for the final content of this report rests entirely with the authoring committee and the institution.

Contents

APPENDIXES

1[1]

Introduction

Glutamate dysfunction has been associated with a wide array of nervous system diseases and disorders. Glutamate-related disorders include neuropsychiatric disorders such as schizophrenia, neurodegenerative diseases such as Alzheimer's, substance abuse, pain disorders, and traumatic brain and spinal cord injuries. These conditions are widespread, affecting a large portion of the U.S. population, and remain difficult to treat (Narrow et al., 2002; Wang and Ding, 2008; Writer and Schillerstrom, 2009). Glutamate's contribution to such a wide range of nervous system disorders is best explained by a single fact: Glutamate is the most pervasive neurotransmitter in the central nervous system (CNS). Despite this fact, no validated biological markers, or biomarkers, currently exist for measuring glutamate pathology in CNS disorders or injuries. A workshop titled *Glutamate-Related Biomarkers in Drug Development for Disorders of the Nervous System* was convened by the Institute of Medicine Forum on Neuroscience and Nervous System Disorders to explore promising current and emerging technologies with potential as reliable glutamate biomarkers, and to outline strategies to

[1] This workshop was organized by an independent planning committee whose role was limited to the identification of topics and speakers. This workshop summary was prepared by the rapporteurs as a factual summary of the presentations and discussions that took place at the workshop. Statements, recommendations, and opinions expressed are those of individual presenters and participants, and are not necessarily endorsed or verified by the forum or The National Academies, and should not be construed as reflecting any group consensus. Furthermore, although the current affiliations of speakers and panelists are noted in the report, many qualified their comments as being based on personal experience over the course of a career, and not being presented formally on behalf of their organization (unless specifically noted).

accelerate development, validation, and implementation of these biomarkers as powerful tools to advance drug development for nervous system disorders associated with glutamatergic dysfunction.

Although glutamate has a staggering array of functions, none of the top-selling CNS drugs is indicated directly for rectifying dysfunction at the glutamate synapse. Currently, three prescription drugs approved by the Food and Drug Administration (FDA)—memantine, ketamine, and D-cycloserine[2]—have implications for diseases of glutamate or glutamate-related pathology. Many workshop participants agreed that the lack of glutamate biomarkers is the largest obstacle to increasing glutamate-specific drug development. In spite of this problem, scientific progress is close to a tipping point that will yield novel glutamate biomarkers as long as concerted efforts are undertaken by academic, government, and industry researchers, as well as by health policy makers. The stakes, in their view, are too great to disregard.

GLUTAMATE BIOMARKERS

Biomarkers are defined as quantitative measurements that provide information about biological processes, a disease state, or response to treatment (IOM, 2008). Although many biomarkers are being investigated expressly for glutamate neurotransmission, none to date has been validated for use in clinical trials, much less clinical practice. The development and adoption of glutamate or glutamate-related biomarkers (hereinafter called "glutamate biomarkers") is crucial because biomarkers streamline research and development of new therapies that have the potential to increase understanding of glutamate-related disorders and make them easier to prevent and/or treat. Daniel Javitt, director of schizophrenia research at the Nathan Kline Institute for Psychiatric Research and cochair of the workshop, stressed that development of glutamate biomarkers has the potential to increase the understanding of glutamate dysfunction in CNS disease.

Biomarkers would provide a mechanism to:

- monitor response to treatment;
- identify people at risk for disease;
- measure and predict disease progression or prognosis;
- identify molecules sufficiently important to the disease that they are strong targets for treatment or prevention;
- offer a choice of outcome measures in "proof-of-concept" studies to spur investment and lead to larger clinical trials;

---

[2] Memantine is indicated for cognitive dysfunction in Alzheimer's disease, ketamine is an anesthetic, and D-cycloserine is indicated for anxiety disorders, schizophrenia, and chronic pain.

- provide mechanisms for patient stratification; and
- find surrogate outcome measures to shorten the length of clinical trials.

To find a potential biomarker, research first must demonstrate that the biomarker has the capacity to reliably distinguish between healthy individuals and those with disease. The process begins with an array of studies, depending on the nature and application of the biomarker and later with replication by other laboratories, all of which require years of investigation. A "potential" biomarker, however, is not automatically designated a "validated" biomarker. The process of validation, for regulatory purposes and thus for clinical trials, requires even more types of studies and various levels of evidence, depending on the use of the biomarker (e.g., measuring drug outcomes) and other FDA requirements (FDA, 2004, 2010; Goodsaid and Frueh, 2007).

WORKSHOP GOALS

In June 2010, the Forum hosted a workshop that examined the potential for development of glutamate biomarkers, and explored next steps that would advance drug development. Established in 2006, the Forum aims to foster dialogue among a broad range of stakeholders—practitioners, policy makers, private industry, community members, academics, and others—and to provide these stakeholders with opportunities to tackle issues of mutual interest and concern. The Forum's neutral venue provides a place for broad-ranging discussions that can help in the coordination and cooperation of all stakeholders to enhance understanding of neuroscience and nervous system disorders. This workshop featured more than 20 presentations describing new approaches to biomarker development while recognizing that the research remains in the hypothesis testing and replication stages.

Specific objectives of the workshop were as follows:

- Briefly outline the need for glutamate-related biomarkers both for understanding the causes of neuropsychiatric disorders and neurodegenerative diseases associated with glutamatergic dysfunction and for accelerating drug development for these disorders.
- Discuss the most promising current and emerging technologies and analytical methods for assessing glutamatergic neurotransmission, and identify the research gaps for their development into biomarkers.
- Outline approaches for biomarker validation in pre-clinical and clinical studies, including relevant animal models and translational challenges.

- Discuss the implementation and regulatory barriers to incorporating glutamatergic biomarkers into drug development for neuropsychiatric disorders and neurodegenerative diseases and approaches to overcome them.
- Identify the next steps in establishing principles and procedures to accelerate biomarker development, validation, and implementation in clinical trials, including frameworks for partnerships and collaboration.

The report that follows highlights the presentations by the expert panelists, and the open panel discussions that took place during the workshop. This report is not intended to be a thorough review of all published literature but an accounting of speaker presentations and commentary by panelists and workshop attendees.

2

Overview of the Glutamatergic System

Glutamate is the major excitatory neurotransmitter in the nervous system. Glutamate pathways are linked to many other neurotransmitter pathways, and glutamate receptors are found throughout the brain and spinal cord in neurons and glia. As an amino acid and neurotransmitter, glutamate has a large array of normal physiological functions. Consequently, glutamate dysfunction has profound effects both in disease and injury.

At least 30 proteins at, or near, the glutamate synapse control or modulate neuronal excitability, noted Darryle Schoepp, senior vice president of neuroscience at Merck. These proteins are membrane-bound receptor or transporter proteins (Figure 2-1). They are strategically situated on several cell types converging on the glutamate synapse: pre- and post-synaptic neurons, astrocytes (a type of glial cell), and nearby inhibitory neurons that use γ-Aminobutyric acid (GABA). GABA is the chief inhibitory neurotransmitter in the brain, and the major difference between glutamate and GABA is that the latter is synthesized from the former by the enzyme L-glutamic acid decarboxylase. Schoepp said the fact that GABA neurons and glutamate neurons are distinguished by this single enzyme could be an efficient way, in evolutionary terms, to control excitability in the nervous system.

Glutamate concentrations in the extracellular space are low and tightly controlled by a large number of mechanisms at the synapse. Perturbations to this regulatory system can have deleterious effects such as excess release of glutamate, which can induce hyperexcitability in post-synaptic neurons to the point of excitotoxicity and cell death (cytotoxicity) (Choi, 1994; Doble, 1999). Glutamate-induced excitotoxicity, particularly in the hippocampus, has been linked to decreased neuronal regeneration and dendritic

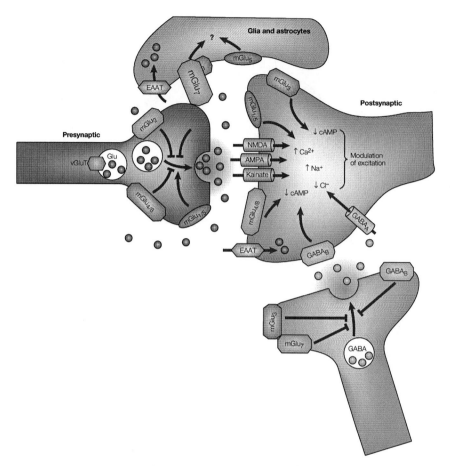

FIGURE 2-1 Illustration of a hypothetical synapse showing localization and function of glutamatergic receptors and transporters.
SOURCE: Swanson et al., 2005.

branching, leading to impaired spatial learning (Cortese and Luan Phan, 2005). Disruptions of glutamate uptake from the synapse have been linked to reduced sensitivity to reward, a symptom of depression (Bechtholt-Gompf et al., 2010). For these and other reasons, a neurotransmitter of glutamate's functional significance must be tightly regulated (Swanson et al., 2005).

The complexity of regulating glutamate and its pervasive presence throughout the brain may explain why, over the past decades, only three prescription medications have been developed that specifically target glutamate or glutamate receptors, memantine, ketamine, and D-cylcoserine. The potential for side effects from these medications is extremely high, which in part deters further investment. By contrast, a broad range of drugs have been marketed to modulate other neurotransmitters, such as dopamine, serotonin, and acetylcholine, whose synaptic regulation is less complex and whose roles and pathways in central nervous system (CNS) pathways are not as extensive.

This presents the most fundamental obstacle facing glutamate bio-marker development and therapeutics: Any agonist or antagonist has the potential to produce beneficial as well as toxic side effects, depending on its concentration, route of administration, dose-related adverse effects, and other key factors. In terms of drug development, the goal is to carefully select a molecular target that modulates dysfunctional glutamate pathways, without disruption of healthy pathways, and minimizes adverse effects. That challenge to glutamate diagnostics and therapeutics was clearly articulated at the outset of the workshop by presenters Schoepp and Dan Javitt, program director in cognitive neuroscience and schizophrenia at the Nathan Kline Institute for Psychiatric Research:

- How can glutamatergic synaptic transmission be selectively modulated in the central nervous system?
- Can we selectively target pathological processes involving the glutamate system?
- Can we monitor the long-term effects of single-target interventions because chronic dosing of any medicine in a system as highly plastic as the glutamate system may not sustain the beneficial effects?

GLUTAMATE RECEPTORS

Glutamate receptors are numerous and highly complex; more than 20 glutamate receptors have been identified in the mammalian central nervous system. They fall into two main categories, ionotropic (voltage sensitive) and metabotropic (ligand sensitive). Each ionotropic or metabotropic receptor has three types, depending on binding specificity, ion permeability, conductance properties, and other factors. Each type has multiple subtypes (Table 2-1). Ionotropic receptors are fast acting and, once opened, can produce large changes in current flow even if the voltage difference across the membrane is small. After glutamate, as a ligand, binds to an ionotropic receptor, the receptor's channel undergoes a conformational change to allow an immediate influx of extracellular sodium among other ions and an efflux of potassium ions. This triggers membrane depolarization in the post-synaptic cell sufficient to induce signal transmission. One of the main glutamate ionotropic receptors, N-methyl D-aspartate (NMDA), is permeable to calcium ions in addition to sodium and potassium ions; calcium ions have both beneficial and toxic effects. The NMDA receptor is unusual because it is a coincidence detector; for the channel to open, glutamate must

TABLE 2-1 Glutamate Receptor Protein Subunit Composition and Properties

Receptor	Protein Subunit	Receptor Properties
Ionotropic Receptors		
NMDAR	NR1, NR2A*, NR2B*, NR2C, NR2D, NR3A, and NR3B	Heterotetrameric; calcium permeability high; long channel open time
AMPAR	$GluR_1$*, $GluR_2$ edited, $GluR_2$, $GluR_3$*, and $GluR_4$*	Heterotetrameric; calcium permeability low if edited $GluR_2$, otherwise moderate; short channel open time
Kainate receptor	$GluR_5$*, $GluR_6$, $GluR_7$, KA1, and KA2	Homotetrameric or heterotetrameric; calcium permeability low; short channel open time
Metabotropic Receptors		
Group 1	$mGluR_1$* and $mGluR_5$	Homodimeric; signals via phospholipase C; located post-synaptically
Group 2	$mGluR_2$ and $mGluR_3$	Homodimeric; signals via adenylyl cyclase; located mostly pre-synaptically; agonists and antagonists mostly distinct from Group 3
Group 3	$mGluR_4$, $mGluR_6$, $mGluR_7$, and $mGluR_8$	Homodimeric; signals via adenylyl cyclase; located mostly pre-synaptically; agonists and antagonists mostly distinct from Group 2

*Glutamate receptor protein subunits for which human autoantibodies have been reported.
SOURCES: Kew and Kemp, 2005; Pleasure, 2008.

bind to the receptor and the post-synaptic cell must be depolarized because the channel is blocked by magnesium at physiological levels and only opens when the cell is depolarized.

Ionotropic receptor channels are formed by assemblies of heterotetrameric or homotetrameric protein subunits. The three types of ionotropic receptors are named after the ligand that expressly binds to one, but not to the other two: NMDA, α-amino-3-hydroxy-5-methyl-4-isoxazolepropionic acid (AMPA), and kainic acid. Once these ligands were discovered, many others, whether agonists or antagonists, were subsequently found (Lesage and Steckler, 2010). Although their properties differ somewhat, as do their anatomical distribution, glutamate receptors are best known for mediating glutamate's role in learning and memory through plasticity, or modification, of channel properties; enhanced glutamate neurotransmission; and gene expression (Barco et al., 2006). Not only are NMDA receptors highly expressed on neurons, but they are also expressed on astrocytes (Lee et al., 2010). The human brain's expansive capacity for plasticity, learning, memory, and recovery from injury is attributed to improvement in synaptic anatomy and physiology of NMDA signaling, most notably in the hippocampus and other regions of the mammalian CNS (Barco et al., 2006). The basic mechanisms underlying plasticity include neurogenesis, activity-dependent refinement of synaptic strength, and pruning of synapses.

Metabotropic glutamate receptors are slower acting; they exert their effects indirectly, typically through gene expression and protein synthesis. Those effects are often to enhance the excitability of glutamate cells, to regulate the degree of neurotransmission, and to contribute to synaptic plasticity (Lesage and Steckler, 2010). Once glutamate binds with a metabotropic receptor, the binding activates a post-synaptic membrane-bound G-protein, which, in turn, triggers a second messenger system that opens a membrane channel for signal transmission. The activation of the protein also triggers functional changes in the cytoplasm, culminating in gene expression and protein synthesis. There are three broad groups of glutamate metabotropic receptors, distinguished by their pharmacological and signal transduction properties. Altogether, a total of eight metabotropic glutamate receptor subtypes have been cloned thus far.

Group I metabotropic receptors are largely expressed on the post-synaptic membrane. They have been implicated in problems with learning and memory, addiction, motor regulation, and Fragile X syndrome (Niswender and Conn, 2010). Group II metabotropic receptors are situated not only on post-synaptic cells, but also on pre-synaptic cells, possibly to suppress glutamate transmission (Swanson et al., 2005). Their dual location may enable them to exert a greater degree of modulation of glutamate signaling (Lesage and Steckler, 2010). Dysfunction of group II metabotropic receptors have been implicated in anxiety, schizophrenia, and Alzheimer's disease. Group III metabotropic receptors, like group II, are pre-synaptic and inhibit neurotransmitter release. They are found within the hippocampus and hypothalamus and may play a role in Parkinson's disease and anxiety disorders (Swanson et al., 2005).

GLUTAMATE TRANSPORTERS

Tight regulation of extracellular glutamate concentrations at both the synapse and in extra-synpatic locations is critical for normal synaptic transmission and to prevent excitotoxicity. Glutamate transporters regulate glutamate concentrations and are situated on both pre- and post-synaptic neurons as well as on surrounding astrocytes, a type of glial cell (Kanai et al., 1994). Five excitatory amino acid transporters (EAATs), previously known as glutamate transporters, have been cloned: EAAT-1 to EAAT-5, with EAAT-2 expressed predominantly on cells in brain regions rich in glutamate (Eulenburg and Gomeza, 2010). It is widely accepted that glutamate transporters on glial cells are primarily responsible for maintaining extracellular glutamate concentrations. However, the presence of transporters on multiple cell types suggests a high level of cooperation (Eulenburg and Gomeza, 2010; Foran and Trotti, 2009; Tanaka, 2000).

Glial cells, most often astrocytes but also microglia and oligodendrocytes (Olive, 2009), perform a key role in modulating extracellular glutamate levels. Under normal conditions, glutamate is recycled continuously between neurons and glia in what is known as the glutamate–glutamine cycle. Excess glutamate in the synapse is taken up by glial cells via EAAT transporters, where it is converted to glutamine. Glutamine is then transported back into neurons, where it is reconverted to glutamate (Rothman et al., 2003). However, glial cells, under certain conditions, may also release glutamate by at least six mechanisms, one of which is reversal of uptake by glutamate transporters (Malarkey and Parpura, 2008). This kind of reverse transport may be involved in brain damage and stroke (Grewer et al., 2008). Finally, altered expression of EAAT-2 is found in amyotrophic lateral sclerosis. The disease is also marked by excess glutamate levels in the cerebral spinal fluid (Rothman et al., 2003). New therapies are being developed to interfere with this pathological process by targeting EAAT-2 on astrocytes (Rothstein et al., 1992).

Given the number of receptors and transporters, the range of cell types expressing them, the variety of regulatory controls, and the narrow concentration difference between normal synaptic function and excitotoxicity, many fundamental questions remain about how to choose potential pharmacological targets. One presenter at the workshop, Schoepp, raised a series of questions and concerns that addressed both biomarkers and choices of molecular target. The foremost concern was whether the biomarker could distinguish between normal physiology versus pathology. What kind of feedback mechanisms and crosstalk at the synaptic cleft must be considered, especially in light of the probable need for chronic dosing of any new medication? Long-term use of any medication might produce unexpected changes, with the potential for side effects or paradoxical effects. The greatest danger is the specter of any glutamate-related drug inducing excitotoxicity and its ramifications.

3

Glutamate Biomarkers

Biomarkers can be categorized in a variety of ways. The most common categorization is by use, such as biomarkers of etiology, pathogenesis, diagnosis, diagnostic subtype, treatment, susceptibility, and progression of disease (FDA, 2010). Biomarkers of use also can extend to the regulatory and public health arena, where they are categorized as surrogate endpoints for Food and Drug Administration approval, biomarkers for clinical practice, clinical practice guidelines, and public health practice (IOM, 2008). Because no glutamate biomarkers are currently validated, this workshop focused primarily on research in the early stages of development. The research discussed ranged from molecular to behavioral. Their applications were primarily aimed at pathophysiology, diagnosis, and treatment. Yet the categorization of biomarkers, posed to participants in the earlier presentations, was more fundamental: What is the best way to conceptualize biomarkers?

Two speakers presented overlapping conceptualizations. The first, by Jeffrey Conn, professor of pharmacology at Vanderbilt University, divided glutamate biomarkers into three general types: (1) biomarkers of structural engagement with a molecular target; (2) biomarkers of functional engagement with a molecular target; and (3) biomarkers of efficacy. The conceptualization by Kalpana Merchant, chief scientific officer at Eli Lilly and Company, divided biomarkers into two types: (1) "proof of mechanism," which includes both target engagement and target modulation based on pharmacodynamic markers; and (2) "proof of concept," which include biomarkers that allow prediction of efficacy or safety.

BIOMARKERS OF ENGAGEMENT AND EFFICACY

A biomarker of structural engagement seeks to answer the seminal question: Does the biomarker bind to, or in any way directly engage with, the molecular target of interest, such as a receptor on a glutamate neuron, a nearby cell, or a transporter? Structural engagement is the rate-limiting step for developing and validating any potential glutamate biomarker. Without demonstrating target engagement, Merchant suggested, it is difficult to determine whether any apparent biomarker is associated with or attributable to the treatment or the underlying disease.

Finding a structural biomarker is a formidable task. It must directly measure factors such as penetration into the brain, degree of receptor occupancy, or another type of direct interaction with the intended molecular target in the central nervous system (CNS). One of the biggest hurdles is penetration into the CNS. The blood–brain barrier (BBB) can prevent penetration or actively extrude certain molecules, often large molecules, by brain reflux transporters. The CNS also can metabolize certain molecules after successfully penetrating the BBB, but before reaching their target (Pike, 2009).

Structural and/or functional biomarkers are best studied by imaging with positron emission tomography (PET), a technology used to detect functional activity in regions of the brain in real-time based on radiotracer ligands binding to targets, in this case at the glutamate synapse. PET affords anatomical and quantitative measurement of displacement of a high-affinity endogenous ligand with a labeled one known as a probe. PET provides information about kinetics with high sensitivity and can map whether the probe can fully occupy a receptor once a sufficient dose reaches the CNS. Dosing information is important because an otherwise excellent biomarker can fail to be identified inside the brain if the dose is insufficient. PET displacement studies of potential biomarkers binding with endogenous ligands, however, do not specify the functional aspects of the probe's interaction with the target. For that purpose, PET can be combined with other imaging techniques, electrophysiology, or another biomarker. Combining techniques, many participants said, may yield more progress than any single technique alone.

A functional biomarker provides a direct measure of target engagement or an indirect measure of downstream actions following target engagement. The biomarker may come from functional magnetic resonance imaging (fMRI), electrophysiological, or electroencephalogram (EEG) response. It could also arise from pharmacodynamic studies using biochemical, physiological, and multimodal imaging techniques. But these and other functional methods must be understood as secondary because they do not directly assess structural target engagement, the primary goal. If a signal appears from one of these functional assays, there is greater confidence that the target is engaged at some level. However, target engagement is not directly assessed except, for example, through a dose–response curve and detailed pharmacokinetic and pharmacodynamic studies. Conn indicated that biomarkers of functional engagement have the strong advantage of providing insight into whether or not a compound may demonstrate or predict efficacy. Functional biomarkers are the most common type of glutamate biomarker currently being studied.

Other types of biomarkers that are important to develop are pharmacodynamic-based measures to stratify diseased individuals based on their response to a given drug. Mark Bear, Picower Professor of Neuroscience at the Massachusetts Institute of Technology, indicated that biomarkers for diagnostic stratification are urgently needed. As with any biomarker for any disease, the biomarker should be minimally invasive.

The range of biomarkers presented below is grouped by modality, that is, the methodological tools used to identify a purported biomarker of structural or functional engagement with a glutamate-related target. Biomarker measurement tools are used for many purposes. They can encompass electrophysiology, genomics, pharmacological response, receptor expression patterns, radiological or other imaging, and behavioral or neuropsychological testing, among others. The methods can be used as direct or indirect measures of glutamate transmission. Any single method or group of methods can be used to shed light on structural or functional engagement, whether at the level of genes, proteins, cells, neurocircuits, cognition, or complex behavior. The methods can be used alone or in combination.

PHYSIOLOGICAL BIOMARKERS

Several functional glutamate biomarkers have gained currency from years of electrophysiological research providing new avenues of research. For example, until recently dopamine transmission was believed to be primarily responsible for the pathophysiology of schizophrenia. However, electrophysiological studies of schizophrenia and the N-methyl D-aspartate (NMDA) receptor have demonstrated that glutamate dysfunction participates in its pathogenesis (Javitt et al., 1996; Umbricht et al., 2000;). In particular, multiple workshop presentations highlighted advancements in electrophysiology techniques as potential methods by which biomarkers for diseases and disorders with glutamate pathology might be developed.

Event-Related Potentials

The brain's processing of sensory information has been studied with electrophysiological techniques for several decades. The human environment is rich with sensory information, much of which is filtered out for its irrelevancy. By contrast, the human brain must be attentive to novel or salient sensory stimuli because that information is evolutionarily crucial to trigger the flight-or-fight response needed for survival. An "event-related potential" (ERP), studied with EEG electrodes on the surface of the scalp, is an umbrella term covering several electrophysiological methods of measuring the CNS response to sensory signals. Several speakers described electrophysiological biomarkers of glutamate dysfunction that rely on specific types of auditory or visual ERPs. ERP biomarkers have been studied in relation to cognitive impairment and negative symptoms in schizophrenia. ERP enables study of schizophrenia impairments not improved by marketed antipsychotic medications. Participants noted that this method helps to understand schizophrenia's cognitive symptoms (e.g., disorganized thinking, inability to plan ahead, impaired memory) or negative symptoms that affect emotion and behavior (e.g., blunted affect, avolition, alogia).

ERP assesses functional areas such as visual and auditory information processing, sensory gating, and slow-wave activity, among others. Some ERP abnormalities manifest early in life, especially in vision and hearing, before onset of schizophrenia. Biomarkers in these sensory systems may help identify those at risk for schizophrenia, as well as monitoring the progression of schizophrenia and measuring drug efficacy for any new drugs targeting negative and cognitive symptoms. For improvements in CNS localization of the source of the response, electrophysiological techniques also can be combined with other techniques, such as magnetoencephalography (MEG), to gain higher spatial resolution of the CNS location generating the signal, as well as with imaging techniques.

Auditory ERPs

Several biomarkers of information processing have links to glutamate dysfunction, including mismatch negativity (MMN) and P3a (Javitt et al., 2008). They are two sequential components of an EEG-recorded waveform, measured by field potentials, which represent the summed synchronous activity of up to millions of neurons. MMN is an ERP measure known as mismatch negativity, a measure of auditory discrimination. It specifically measures the decline in EEG amplitude (in μvolts) in response to an auditory stimulus that is distinct from a stream of otherwise repetitive auditory stimuli (Figure 3-1). The singularly dissimilar stimulus is referred to as the "oddball stimulus." Detected by a scalp electrode above the auditory cortex, the MMN is recorded about 200 milliseconds after the oddball stimulus is introduced. It is shown in the figure below as the nadir in the waveform. It is evoked even when subjects are not told to attend to it.

Several hundred milliseconds after introduction of the oddball stimu-

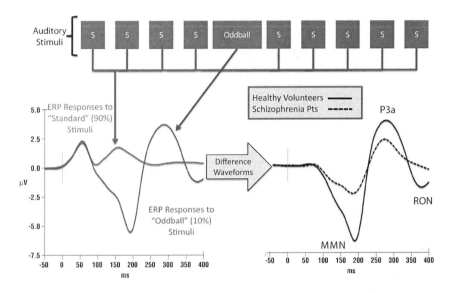

FIGURE 3-1 Auditory biomarkers MMN and P3a from electrophysiology.
SOURCE: Light et al., 2010.

lus, the waveform amplitude surges upward, marking the "P300" peak amplitude, with P signifying positive amplitude at 300 milliseconds. One component of the P300 peak, known as P3a, is measured by electrode placement above the frontocentral lobe. The P3a component is detected by EEG.[1] Thus, the CNS's passive registration of the oddball stimulus is measured first by an amplitude decrement (MMN) prior to an amplitude increase (P3a) in healthy people.

In individuals with schizophrenia, the waveform pattern is noticeably different. The general waveform is similar to that of healthy people, but significantly blunted in negative and then positive amplitude (Figure 3-2). Gregory Light, associate professor at the University of California–San Diego, explained that changes in MMN and P3a are linked to a broad array of other features of schizophrenia, including decrements in higher order cognitive processes, measures of drug efficacy, and patients' daily functioning, among other measures of global assessment. MMN abnormalities have been found to be heritable in people with schizophrenia (Hall et al., 2006a, 2006b). NMDA antagonists reproduce the neurophysiological waveform

---

[1] When the lead is placed at another site on the skull, the so-called P3b component of the ERP is associated with cognitive processing because the subject reports having detected the oddball stimulus at that latency in time.

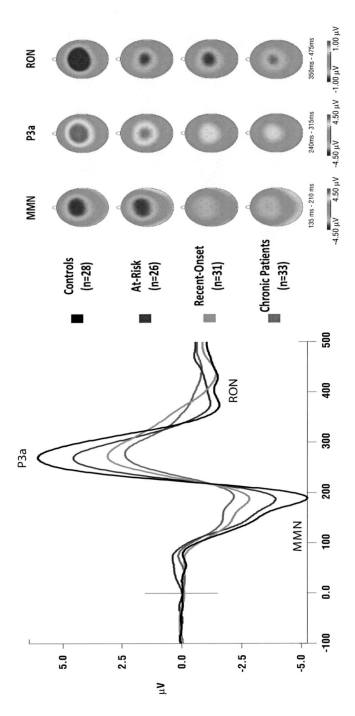

FIGURE 3-2 Automatic sensory information processing abnormalities across illness course of schizophrenia.
SOURCE: Jahshan et al., in press.

of schizophrenia in both animal and human models suggesting that MMN has the potential for use as a technique for glutamate biomarker identification (Javitt et al., 2008). The role of serotonin, dopamine, nicotinic, and other receptors in the generation of MMN is less clear; however recent studies have found that MMN amplitude and latency are altered following antagonist treatments, such as haloperidol and psilocybin (dopamine and serotonin receptor antagonists respectively) and following nicotinic receptor stimulation (Garrido et al., 2009).

Changes in MMN and P3a are consistently replicated abnormalities in schizophrenia, noted Gregory Light, associate adjunct professor at the University of California–San Diego. In a meta-analysis, the effect size of the MMN waveform differential between healthy subjects and those with schizophrenia is approximately one standard deviation (Umbricht and Krljes, 2005). When examining the relationship between the waveform and the course and severity of schizophrenia, the findings are striking: The group "at risk" for schizophrenia tracks the more normal waveform (in P3a pattern), yet as disease progression occurs, the waveform becomes less and less pronounced (Figure 3-2). In healthy subjects, the waveform pattern shows no changes over the course of time. Finally, when given to healthy people, the NMDA antagonist ketamine induces MMN (Umbricht et al., 2000) and P3a (Watson et al., 2009) attenuation similar to that seen in schizophrenia.

While clinical application of ERPs to schizophrenia is promising it has also proven useful for investigating other diseases including dyslexia and learning disabilities (Garrido et al., 2009). Interestingly, the changes to the P3a component are not unique to schizophrenia and have been found in Alzheimer's disease, bipolar disease, and attention-deficit hyperactivity disorder (Javitt et al., 2008).

Steady-State, Visual-Evoked Potentials

The visual system is rich in glutamate neurotransmission from the retina through nuclei en route to the visual cortex. Visual defects are manifest in schizophrenia, affecting about a third of patients (Butler et al., 2005). Brian O'Donnell, professor of psychology at Indiana University, noted that people with schizophrenia display deficits in early-stage processing of visual information by the magnocellular pathway of the visual system, as assessed by steady-state, visual-evoked potentials. The magnocellular pathway transmits visual information of low resolution from the retina through the thalamus to the visual cortex, as opposed to high-resolution information transmitted by the parvocellular pathway.

As measured by psychophysical tests, the deficits include dot-motion trajectory discrimination, grating velocity discrimination, and contrast sensitivity, among others. Although the underlying mechanism is unknown, an impact is seen on the perception of motion/spatial processing as opposed to object processing. Steady-state, visual-evoked potentials consist of presenting a stimulus that is periodically varying, such as a flickering patch or grading on a screen. People with schizophrenia display a selective reduction in these steady-state evoked potentials, with deficits greatest at frequency bands of 17 Hz and higher (Krishnan et al., 2005). There is also a visual amplitude drop in the P300 test, described earlier, but the effect is not as strong as that seen in the auditory system, O'Donnell stated. Visual deficits are correlated with problems in independent living scales, one component of measurement on the Global Assessment of Functioning scale (Butler et al., 2005). Although only tested in normal animals, not humans, the NMDA antagonist ketamine impairs discrimination of horizontal and vertical lines formed by spatial proximity of dots (Kurylo and Gazes, 2008). Although the animals discriminated solid patterns normally, they performed abnormally on the type of visual deficits found in people with schizophrenia. O'Donnell suggested that because the visual system offers noninvasive access to the CNS and perceptual defects are associated with glutamate dysfunction, biomarker development in this area is worth pursuing.

Sensory Gating

Sensory gating is another type of biomarker obtained by EEG. It refers to an automatic process that enables the brain to adjust and habituate to a series of repeating sensory stimuli. After an initial stimulus, the brain suppresses its response to a repeated presentation of the same stimulus. Two types of auditory gating measures P50 and prepulse inhibition (PPI) were described by Mark Geyer, professor at the University of California–San Diego, and Bruce Turetsky, associate professor at the University of Pennsylvania, as potential glutamate biomarkers. These two measures have been studied in animals and humans with schizophrenia, among other disorders. Both measures are abnormal in schizophrenia, suggesting potential biomarkers of pathophysiology. Although both are widely recognized biomarkers, neither is specific to schizophrenia; P50 sensory gating is abnormal in Alzheimer's disease while both are abnormal in bipolar disorder. P50 is also found abnormal in yet another disorder, cocaine abuse (Javitt et al., 2008). The two measures are neither strongly correlated with each other, nor with cognitive abnormalities in schizophrenia (Greenwood et al., 2007). P50 does not measure glutamate function if the latter is defined by response to ketamine.[2] Turetsky indicated that ketamine has no effect on

---

[2] Ketamine is a non-competitive antagonist that blocks the NMDA receptor channel. Other competitive agonists and antagonists for the NMDA receptor might be different ways to assess glutamate function or dysfunction. See the final section of this workshop report.

the P50 gating response in both rodents and human volunteers. In contrast, the pharmacological evidence linking PPI and glutamate is stronger. Geyer reported that NMDA antagonists disrupt PPI in rodents, whereas clozapine, a widely used drug for schizophrenia, prevents the disruption. The animal data suggest that PPI could be used as a biomarker in animal models for pharmacological studies of glutamate-related medications, but the applicability to humans is not completely clear.

EEG/MEG Combination

A combination of EEG and MEG is being tested as a potential biomarker of depression and its early response to treatment (Tononi and Cirelli, 2006). Depression is one of the leading causes of worldwide disability (WHO, 2001). Most antidepressants fall under the umbrella of serotonin- and norepinephrine-targeted drugs and, for bipolar depression, anticonvulsants or antipsychotics. The drugs usually take several weeks to achieve full effects. Given these drawbacks, glutamate has been studied as another target for modulating depression, possibly with more rapid effects. Glutamate pathways appear to contribute or modulate depression in animal models and humans, as shown in studies using MRI and post-mortem tissue analysis. α-Amino-3-hydroxy-5-methyl-4-isoxazolepropionic acid (AMPA) receptor potentiators of synaptic plasticity and an NMDA antagonist both displayed rapid antidepressant effects in clinical trials (Brennan et al., 2010; Sanacora et al., 2008; Zarate et al., 2006, 2010).

To analyze the role of glutamate pathways and antidepressant effects, Carlos Zarate, chief of the mood and anxiety disorders research unit at the National Institute of Mental Health, reported that studies are being conducting in depressed patients that have failed most treatments, such as lithium and quetiapine, with a combination of high-density EEG and MEG activity to analyze slow-wave activity above the anterior cingulate region. High-density EEG uses multichannel recordings to localize neurocircuits arising from *extracellular* currents, and MEG[3] is used to identify *intracellular* currents. The union of techniques has allowed Zarate and colleagues to predict which patients will show the greatest improvement in symptoms following administration of a low-dose of ketamine. In these patients, ketamine was found to have a significant rapid antidepressive effect, within 110 minutes of administration (Salvadore et al., 2010; Zarate et al., 2006). The union of techniques can help to localize, in this particular context,

---

[3] MEG records the magnetic fields engendered by electrical currents within the brain. It uses arrays of superconducting quantum interference devices. The main application is to localize pathological reasons prior to surgical removal, neural feedback, and research (with the last goal to determine the function of various parts of the brain).

changes in synaptic potentiation as evidence by increases in slow-wave activity. Data from this preclinical study suggests that the effects of ketamine are immediate through directly targeting the NMDA receptor complex which indirectly enhances the AMPA throughput, leading to rapid antidepressant effects. In summary, these newly combined electrophysiological measures may serve as a biomarker to predict drug efficacy.

Animal Models

The basis of many electrophysiological findings, which have been developed in humans, may be correlated with positive outcomes in disease treatment, but they may not be fully understood. Animal models offer insight about molecular, cellular, and neurocircuits that underlie electrophysiological findings. They can help to clarify the mechanisms of signal transduction, the cell populations most responsible for generating signaling, and the circuitry responsible for neural oscillations, and they can enable testing of new drugs. One presentation addressed the value of primate models in investigating ERPs, such as MMN and other noteworthy electrophysiological biomarkers.

Schroeder and his team of investigators have developed the use of multiarray electrodes recording at different lamina within the auditory and visual cortices of non-human primates. Studying selective attention, they have sought to localize the source of the oscillations responsible for selective attention in visual and auditory tasks, and to interpret its causation. The best method they have found to understand the ERPs is by examining a particular parameter known as current-source density (Lakatos et al., 2008). Applying that technique, Charles Schroeder at the Nathan Kline Institute has reported cell type; cell population (e.g., pyramidal cells); the pattern of circuit activation (e.g., feedforward, feedback); the physiological identity of transmembrane currents; and net local excitation versus inhibition. These and other studies signify the importance of animal testing in biomarker development to find applications to humans, many participants noted.

COGNITIVE BIOMARKERS

Cognition represents a high-functioning capability of the brain, covering areas such as memory, language, planning, abstract reasoning, and reasoning speed. Cognition is measured by neuropsychological tests, many of which have been validated for purposes of studying cognitive impairment (e.g., impairment in psychiatric disorders, substance dependence, and memory impairment in Alzheimer's disease). The tests are delivered by questionnaires and/or verbal examination administered by professionals.

One cardinal form of cognition is a psychological construct known as "working memory." It is carried out in the dorsolateral region of the prefrontal cortex of the brain in conjunction with neurons from the parietal, temporal, and cingulate cortices (Friedman and Goldman-Rakic, 1994). Working memory refers to the ability to maintain and manipulate information over a short period of time, usually 30 seconds, in order to plan, solve problems, and reason capacities collectively described as goal-oriented behavior. Glutamate likely mediates these functions (Lewis et al., 2003). Serious problems in working memory are responsible for many symptoms of schizophrenia (Goldman-Rakic, 1994; Verrall et al., 2010). These symptoms are difficult to treat and responsible for the most disabling characteristics of schizophrenia (Buchanan et al., 2005).

One major psychological test of working memory function is the AX-Continuous Performance Task (CPT) (Barch et al., 2009). The task requires subjects to press a button when they pick out the letter "X" preceded by the letter "A," amid many other letters preceding the "X." Performance on this task, which measures goal maintenance, is abnormal in people with schizophrenia. In healthy people, the NMDA antagonist ketamine produces schizophrenia-like deficits on the AX-CPT (Umbricht et al., 2000). While AX-CPT may be sensitive to other neurotransmitter antagonists, such as muscarinic receptor antagonists, the ketamine challenge finding suggests that NMDA receptors are dysfunctional in schizophrenia.

Another working memory task is known as "n-back." In this task, the volunteer is required to follow a series of stimuli and is instructed to respond whenever a stimulus is presented that is the same as the one previously introduced n trials ago, wherein n is a pre-specified integer, most typically 1, 2, or 3.

Conducting a large study of more than 1,000 healthy individuals, Angus MacDonald and his colleagues at the University of Minnesota found that those screening positive for D-amino oxidase (DAO) single nucleotide polymorphisms (SNPs) showed significantly more errors on the AX-CPT and the n-back tasks. DAO metabolizes the amino acid D-serine, which is a coagonist of the NMDA receptor (Verrall et al., 2010). This finding suggests a potential link between DAO SNPs and schizophrenia, as the disease has been linked to poor working memory. Still, more work needs to be done to establish AX-CPT and n-back neuropsychological tests as potential biomarkers for schizophrenia and/or glutamate dysfunction. One problem is that poor performance on these cognitive tests is not specific to schizophrenia (Javitt et al., 2008). MacDonald explained that these psychological tests have not been studied sufficiently in large samples and by other methods of psychometric validation.

IMAGING BIOMARKERS

Imaging biomarkers have flourished over recent decades as a result of striking advances in technology. In fact, MRI and computed tomography (CT) scans are frequently named in physician surveys as among the foremost advances in medicine during the 20th century. Both are now standard diagnostic tools in medical practice. But many newer imaging techniques are used in research and have not yet been adopted in clinical practice, largely because of a combination of high costs and the high level of sophistication required for proper use. Imaging can also be used with other methods to corroborate or refute findings. Knowledge gained through these methods can streamline biomarker development and drug development, said most workshop presenters.

Many presenters described a variety of imaging techniques used in the pursuit of glutamate biomarkers. Roughly, the techniques described below deal with structure, function, and/or quantification of functional activity. Function can be objectively measured, often by signal density, intensity, duration, and location. Imaging can be used alone or in combination with electrophysiology. As a general rule, electrophysiology is better suited to detect functional changes within a time frame of milliseconds. This is especially vital considering that action potentials take up to 130 ms to propagate. Imaging is superior for spatial resolution, but its time resolution typically ranges from seconds to minutes. Both have the capacity to measure functional activity. So imaging is generally superior for spatial resolution, whereas electrophysiology is superior for temporal resolution.

MRI

MRI is ideally poised to study in vivo brain structure and function. The use of non-ionizing radio frequency signals for image acquisition is tailored to soft tissue and safer than CT, which exposes the patient to high levels of radiation. fMRI and pharmacological MRI (phMRI) are broadening the understanding of glutamate sites of action, mechanisms of action and treatment localization, dose, and effects. fMRI and phMRI depend on changes in the amount of blood oxygenation level–dependent (BOLD) signal. Lacking reserves for glucose and oxygen, active neurons require more immediate glucose and oxygen delivery than do inactive neurons. A BOLD perturbation in hemodynamic activity is associated with higher activity in nervous system pathways and regions under study. fMRI is frequently undertaken during execution of specific tasks assigned to subjects, such as neuropsychological tests measuring cognition, emotion, or substance dependence.

A union of pharmacological techniques and fMRI, phMRI is used to detect drug-induced effects on activity levels in the brain at the site of action. Drug administration is given to examine signaling changes from baseline. The technique seeks many answers, such as the location of the drug's physiological target, the response to treatment, appropriate dose levels of a drug to achieve a desired effect, and the pharmacokinetics and pharmacodynamics at its site of action (Paulus et al., 2005). Activity-dependent changes with phMRI also may identify the consequences of drug action over time, its distribution to other areas of the brain, and aspects of pathophysiology, disease onset, and course of disease. phMRI is especially useful for studying glutamate because glutamate's action is highly energy dependent and thus readily detectable. Another advantage is that phMRI may identify sites of adverse effects that could be targeted for protection. Finally, an advantage of phMRI over PET is that no radioligands are necessary to find and produce, which is the most difficult and sophisticated part of the process.

Magnetic Resonance Spectroscopy

One variant of magnetic resonance techniques is proton magnetic resonance spectroscopy (^{1}H MRS). It can assess in vivo function of brain chemistry by exploiting the fact that hydrogen atoms in distinct chemical environments, depending on the molecules in which they are bonded, possess distinct resonant properties. Glutamate and glutamine are among the few molecules detectable by MRS, but at extremely low concentrations (6.0–12.5 mmol/kgww). The concentrations can be inferred from the spectra generated by the technique. MRS is spatially and temporally averaged to detect molecules of interest. In his presentation, Robert Mather, principal scientist at Pfizer Pharmaceuticals, described his application of MRS to the study of retigabine, an anticonvulsant drug that binds to voltage-dependent potassium channels (Kv7 or M-channels). MRS enabled him to determine that retigabine reduced concentrations of glutamate/glutamine in the hippocampus. His study demonstrated the feasibility of using MRS as a mechanistic biomarker of changes in glutamate/glutamine ratios associated with the action of a particular medication at its target location. Recent studies using MRS have found abnormalities in the glutamine/glutamate ratio in bipolar disorder and major depressive disorder with reductions during depressive episodes and increases during mania, similar to those seen during the first episode of schizophrenia (Yüksel and Öngür, 2010). Dr. Mather remarked that MRS can be a valuable biomarker of glutamate levels in many psychiatric and neurological diseases.

PET and SPECT

PET is a non-invasive, *in vivo* nuclear medicine imaging technique adept at studying CNS function. PET radioligands, which must be short lived to reduce radiation exposure, are injected intravenously and synthesized to interact with particular molecular targets. PET's foremost strength is its level of sensitivity (10^{-12}M) and capacity for kinetic analysis. Drawbacks are its slower temporal resolution (minutes) and low spatial resolution (2–6 mm). To improve both, PET can be combined with a number of other imaging methods performed simultaneously and built into the same machinery, such as single photon emission computed tomography (SPECT), which furnishes 3-D images that can be manipulated, among other distinctions. The combined techniques provide better understanding of molecular interactions, molecular environment, and understanding of receptor occupancy.

During the first few years of PET's introduction, starting in the 1960s, PET radioligands were limited; the foremost ligands, oxygen-15 and fluorine-18, among others, were too nonspecific for studying glutamate synapses. What also hindered PET's application to glutamate was its limited spatial resolution. Today, however, PET has dramatically improved, as has the introduction of more PET ligands and new techniques to reconstruct PET images for better spatial resolution.

The foremost barrier holding back PET applications for glutamate neurotransmission is the scarcity of PET radioligands that expressly bind to particular molecules at the glutamate synapse. PET remains a challenge because of the complex interplay of 30 or more molecular targets vying for glutamate modulation, observed Schoepp. Other problems specifically with PET ligands have been lipophilicity, which leads to nonspecific retention in the high levels of CNS lipids. Radioligands have also been beset by high binding to plasma proteins, which limits entry of the radioligand to the CNS.

One of the few examples of PET radioligands for target engagement in glutamate neurotransmission is a specific radioligand for the mGluR5 receptor. Speaker Robert Innis, of the National Institute of Mental Health, and his team developed an 18F-labeled ligand for the compound SP203, which antagonizes the mGluR5 receptor. He reported finding that PET scanning could be used to visualize and quantify the labeled antagonist in the healthy human brain. PET provides a level of detail not available with other techniques. MRI could not be used for reconstructing images because its sensitivity was insufficient to detect the potential biomarker. He also found that animal models could not have predicted target engagement in humans because rats and monkeys defluorinated the radioligand before its uptake into the CNS. The glutamate antagonist is highly important to study in vivo because its antagonism of the metabotropic receptor blunts drug-seeking behavior, which is of major significance to the study of drug dependence.

One other radioligand is an 18F-labeled inverse agonist for the cannabinoid receptor CB_1 (Pacher et al., 2006; Terry et al., 2010). The receptor mediates marijuana's psychotropic effects. Located pre-synaptically, CB_1 receptors inhibit the release of glutamate. CB_1 receptors are located widely throughout the brain, including the cortex, hippocampus, and cerebellum. In glutamate-related diseases and injuries, these receptors function pathologically, likely by releasing excess glutamate and producing excitotoxicity and oxidative stress on the post-synaptic neuron (Pacher et al., 2006; Terry et al., 2010).

SPECT and PET are similar techniques except that SPECT directly emits gamma radiation, whereas PET emits two gamma photons in opposite directions. A PET scanner has the distinct advantage of generating significantly higher resolution images, about two to three orders of magnitude greater than SPECT. But SPECT images can be manipulated in three dimensions. To protect human health, the gamma-emitting radiotracers used in PET must be short lived, whereas SPECT uses radiotracers that are longer lived isotopes.

4

Treatment Implications of Biomarkers

Biomarker development has the potential to increase the efficiency of drug development, refine or enhance clinical trial data, and speed access to safe drugs (IOM, 2008). For nervous system disorders, these goals might be accomplished through new target development, patient stratification, and side-effect reduction. This section highlights three central nervous system (CNS) disorders—ischemia, autism spectrum disorders, and chronic pain—where current glutamate biomarker research has the potential to advance drug development.

TARGET DEVELOPMENT

One of the tenets of glutamate dysfunction is that increased extracellular glutamate levels, under conditions of ischemia and trauma, set in motion a cascade of events that lead to intense calcium influx into glutamate neurons. With large participation by astrocytes, calcium influx into post-synaptic glutameteric neurons leads to widespread cell death from excitotoxicity and necrosis (Choi, 1994). Focal ischemia accounts for 80 percent of stroke damage. But this basic tenet of glutamate's dominance is more nuanced, as a result of a decade or more of research.

Dennis Choi, executive vice president at the Simons Foundation, described current thinking about glutamate as having lost momentum for new drug development to block calcium influx by N-methyl D-aspartate (NMDA) receptor blockade. One of the main problems, Choi said, was that animal studies with positive results did not translate to humans. With the benefit of newer research, Choi explained that, contrary to expectations, ischemia is not wholly accountable for neuronal necrosis, nor are excess calcium (Ca^{2+}) ions. Other cations and receptors do play a profound role, suggesting potential new targets for drug development.

Further research has revealed roles for glutamate's metabotropic receptors, as well as other ions and receptors beyond glutamate's. Although more is known, Choi emphasized that the lack of biomarkers has been highly detrimental to progress. Calcium excess in ischemia still triggers necrotic cell death in the first hours, so blocking NMDA receptors during that window is important. But afterward, sufficient calcium release also causes release of zinc (Zn^{2+}) ions into the extracellular space, which, in turn, blocks the NMDA receptor. At that point, NMDA blockade should cease. But there are no biomarkers to determine when that point occurs. Meanwhile, ischemia is known to cause acidosis, with proton release from ATP hydrolysis (Xiong et al., 2008). The pH levels of the brain fall to 6.5. That pH drop activates receptors throughout the brain known as acid-sensing ion channels (ASICs), which are proton-gated cation channels widely distributed in peripheral sensory neurons and the CNS.

Acidosis through activation of ASICs also is responsible for substantial neuronal injury. The greater the acidosis with calcium and sodium ions acting through ASICs, the greater the infarct is (Xiong et al., 2008). Using biomarkers within the CNS, were they to exist, researchers could understand how the movements and timing of zinc, calcium, and hydrogen increases help to mitigate the impact of ischemic stroke. Choi concluded that although glutamate is an important player, biomarkers for ASIC's action and cations are essential to pave the way for new target development.

PATIENT STRATIFICATION

The potential for patient stratification became clear in the discussion of glutamate biomarkers for autism spectrum disorders (ASDs). ASD, as its name implies, covers a broad spectrum of symptoms. Although most genetic causes of autism are unknown, several single-gene disorders are associated with high rates of ASD. The best understood genetic subtype is caused by a genetic mutation in the Fragile X gene, but Fragile X only accounts for 2 to 5 percent of those with ASD (Kelleher and Bear, 2008). Although those with the Fragile X gene have symptoms that overlap with other ASD cases, the underlying basis may be fundamentally different. The next most common single-gene disorder causing autism is tuberous sclerosis complex (TSC).

The single gene defect in Fragile X syndrome silences FMR1, the gene encoding the Fragile X protein, which normally represses protein translation. When silenced by the mutation, Fragile X is responsible for mental retardation (Bagni and Greenough, 2005). The silencing of the FMR1 protein causes excess signaling through mGluR5. The receptor itself is not defective; the defect is in the heightened rate of translation, at the presynaptic terminal, of up to 400 distinct mRNAs that FMR normally represses (Brown et al., 2001). The lack of repression markedly increases the rate of protein synthesis at axonal terminals, including proteins for glutamate signaling. As further confirmation, studies of Fragile X knockout mice found a decrease in mGluR5 signaling reversed the syndrome's manifestations (Dölen and Bear, 2008). These findings led to ongoing clinical trials of several metabotropic glutamate antagonists. But one group of autism patients is expected to be adversely affected by the treatment—those with tuberous sclerosis.

Tuberous sclerosis, an autosomal dominant disorder, is caused either by mutations of hamartion (TSC1) or tuberin (TSC2). Those defective proteins act in the brain to inhibit protein synthesis at axonal terminals. Decreased protein synthesis is the polar opposite of the effects of Fragile X mutation. Consequently, a metabotropic antagonist given to patients with tuberous sclerosis would likely block glutamate signaling to such a great extent that it would be deleterious. The awareness of opposing functions of two genetic causes of autism highlights the need for stratifying patients by genotype. But the vast majority of autism cases have no genotype biomarkers. Developing biomarkers of increased or decreased rates of protein synthesis at synaptic terminals have been largely unsuccessful, stated Mark Bear, Picower Professor of Neuroscience at Massachusetts Institute of Technology.

Research on autism has brought to the fore a major challenge for drug development: predicting outcomes for patients with the same diagnosis, but with different subtypes, noted Bear. Recognition of this challenge began with excitement surrounding novel treatments now being tested in clinical trials. However, Bear indicated, although one group of autism patients has responded well, another group of autism patients, it could be reasoned, might be harmed because of the lack of biomarkers to stratify patients with different subtypes of disease. What has been learned about autism pathophysiology over the past decade has pointed to the importance of stratifying patients with the same diagnosis but a different genotype, such as Fragile X and tuberous sclerosis, to predict treatment outcome.

Most neurological and psychiatric diseases are diagnosed by symptoms, history, and course of illness without the benefit of biomarkers. If more were known, patient subtypes would likely emerge. Grouping patients by subtypes maximizes the opportunity to understand causation and find new treatments. A clinical trial could be designed strictly for individuals with that subtype rather than all patients with the same diagnosis. Lack of homogeneity in a treatment group may decrease the chances of finding a robust effect or even preclude finding an effect, stated William Potter, cochair emeritus of the Neuroscience Steering Committee at the FNIH Biomarkers

Consortium. The drug that may be effective for one subtype may be ineffective or possibly harmful for another. The clinical trial may be stopped prematurely or investment in research might be halted unnecessarily—to the detriment of patients with a particular subtype.

SIDE-EFFECT REDUCTION

Chronic pain syndromes are highly prevalent, affecting up to 30 percent of the U.S. population (Johannes et al., 2010). Of two marketed glutamate-targeted drugs, one (ketamine) is an anesthetic, thereby indicating the prominent role that glutamate plays in pain. The problem with ketamine is that the CNS action is associated with side effects, including confusion, drowsiness, learning and memory impairment, and ataxia. This and the inaccessibility of the CNS and the lack of consensus on what is the host of the sensitization mechanisms are impeding progress. To obviate part of the problem, Brian Cairns, associate professor at the University of British Columbia–Vancouver, raised the importance of looking for pain-related glutamate biomarkers in the peripheral nervous system, where few have looked in any detail. The concept of peripheral sensitization biomarkers arose from his knowledge of the adverse effects of CNS treatments and his growing understanding of pain pathophysiology.

Glutamate is the primary neurotransmitter for sensory neurons carrying pain information from the periphery to the CNS and within the CNS. Much of the difficulty of finding pain biomarkers arises from research pointing to CNS sensitization as the main driver of chronic pain. In many chronic pain syndromes, including temporomandibular disorders (TMDs) with muscular pain, the chronic pain is experienced[1] as a result of plasticity in the form of central sensitization in the somatosensory cortex. Central sensitization can occur after prolonged increase in excitation of pain neurons in the CNS triggered by sustained, repetitive, and high-frequency input from nociceptors (i.e., pain sensory receptors in the periphery). Central pain sensitization is exaggerated pain signaling. It is the pathological equivalent of long-term potentiation. One manifestation of central sensitization is that painful stimuli that would normally cause minor pain instead induce exaggerated pain (hyperalgesia).

Hyperalgesia is manifest in TMD and in many other pain disorders. In animal models, its molecular basis begins with high and prolonged exposure to a painful stimulus, which triggers release of inflammatory cytokines by the immune system. Cytokines and other proinflammatory agents trigger

---

[1] The International Association for the Study of Pain does not define pain as a sensation. Rather, pain is an unpleasant sensory and emotional experience associated with actual or potential tissue damage.

spontaneous discharge of peripheral nociceptors, leading to transmission of pain signals to the CNS and induction of pain behaviors (Cairns, 2010).

NMDA antagonists have been used in research and clinical practice to attenuate pain. Cairns and his colleagues sought to identify an NMDA antagonist that was active primarily in the periphery. NMDA receptors are diheteromeric, consisting of two subunits: two NR2A or two NR2B units. The latter are found preferentially in the periphery (Collingridge et al., 2004). The finding that 40 to 60 percent of masseter nociceptive neurons in the periphery express the NR2B subunit provided the rationale for testing a peripherally acting NMDA antagonist to protect against central side effects of NMDA antagonists (Gazerani et al., 2010). Cairns found that glutamate-evoked masticatory muscle afferent discharge is mediated through peripheral NR2B subunits. In a rodent model, the NR2B antagonist ifenprodil reduced the glutamate-evoked masticatory muscle sensory afferent discharge. In a human trial, ketamine reduced TMD pain approximately one hour after a single injection into the masseter muscle (Castrillon et al., 2008). Research also revealed that patients with muscular pain in TMD were found to have elevated levels of glutamate in their masseter muscles (Castrillon et al., 2010). Direct intramuscular injection of glutamate induced pain that was mediated through activation of peripheral NMDA receptors, Cairns explained.

Taken together, these and other findings are interpreted as supporting peripheral sensitization of nociceptive afferent fibers in TMDs, Cairns said. Peripheral sensitization due to increased glutamate levels in tissues and activated glutamate receptors have been underestimated as part of the pathophysiology of chronic pain. Both peripheral and central sensitizations are likely at play. Based on the studies with ketamine or ifenprodil, he reaffirmed the need to develop and test peripherally restricted NMDA receptor antagonists. Finally, he concluded that elevated interstitial glutamate levels in the periphery might be a potential target that would allow development of drugs that treat pain, but avoid the side effects found in current drugs that target the CNS.

5

Challenges and Opportunities

Despite the fact that glutamate is a key excitatory neurotransmitter which plays an important role in many central nervous system (CNS) disorders, few biomarkers have been developed to provide objective measures of diagnosis, treatment, and/or prognosis for glutamatergic diseases. The few biomarkers tied to glutamate function have not advanced far enough to allow for simple "go/no go" measures of drug effect on glutamate in the brain. Workshop participants highlighted four key features of the glutamatergic system that are hurdles for biomarker research:

- The diversity of glutamate's functions and the ubiquity of its expression in CNS pathways can lead to widespread pathology and greater likelihood for adverse effects with non-specific treatment;
- The plasticity of the glutamate synapse is an important feature of glutamate transmissions, but can be highly damaging and difficult to control in disease states;
- Glutamate physiology versus pathology depends on relatively minor differences in glutamate concentrations; and
- Tight control over normal glutamate neurotransmission is exerted by at least 30 proteins found at the synapse. Targeting one or more of these proteins can trigger feedback mechanisms that might negate the intended effect.

The complexity of this system presents a formidable barrier to developing glutamate biomarkers, making glutamate-related diseases difficult to treat or prevent. However, embedded within the complexity is tremendous opportunity, as shown by identification of countless targets and emerging new scientific techniques.

CHALLENGES

Over the course of the workshop, participants noted several challenges to the development of glutamate-related biomarkers. They include

- Few pharmacological ligands in animals or humans are available to probe the glutamate synapse. Few radiotracers exist for imaging glutamate function, especially through positron emission tomography (PET);
- Findings from animal models often cannot be translated to clinical trials; likewise, modeling of human diseases, especially neuropsychiatric disorders, are difficult in animals;
- The lack of standardized biomarker characterization (e.g., electroencephalogram, event-related potentials, functional magnetic resonance imaging) is slowing cross-site comparisons and testing of biomarkers; and
- A lack of biomarkers that allow for patient stratification, thereby potentially yielding mixed findings in clinical trials.

Still, participants expressed enthusiasm that the science is near a tipping point. With coordinated investment in biomarker development, glutamate-related drugs are closer to realization than ever. The key is to advance understanding in the glutamate system and provide opportunities for drug development that have not yet been realized.

OPPORTUNITIES

Biomarkers hold the key; they are vital to refine targets, provide proof of mechanism, provide proof of concept, and evaluate whether early intervention focused on a single target can prevent or forestall disease. A number of opportunities for advancement were identified over the course of the workshop including the development of disease-specific genetic and epigenetic biomarkers, the development of new animal models, and the study of small molecules, or metabolomics. Here we highlight three opportunities that emerged as important next steps during discussions.

Development of More PET Ligands

PET enables non-invasive assessment of molecular activity in humans and animals *in vivo*. Its premier benefit is to localize and quantify molecular events, including signal transduction, gene expression, and protein–protein interactions (Jacobson et al., 2002). With respect to the glutamate synapse and drug development, the foremost application of PET is to show proof of mechanism, particularly by structural target engagement, pharmacokinetics and pharmacodynamics, dose finding, and/or patient stratification.

Despite the promise of applying PET-based techniques to the molecular level of the glutamate synapse, PET requires short-lived radioactive ligands that are difficult to synthesize. Many ligands have been developed for acetylcholine, dopamine, and benzodiazepine pathways, but fewer for glutamate pathways. The lag in development, and the importance of finding glutamate-related treatments, inspired many participants to emphasize the need for more concentrated efforts on PET ligand development for the glutamate synapse.

In discussions among participants, many voiced the need for more radioligands for highly specific glutamate-related proteins at the synapse. One of the most important is to find a PET ligand for ketamine in order to gain more information about its non-competitive binding site within the N-methyl D-aspartate (NMDA) receptor channel, a site currently unidentified. Ketamine has been studied as a treatment for multiple CNS disorders, including depression and pain. The availability of PET ligands to investigate competitive, as well as noncompetitive, NMDA antagonists is also important, according to Geyer.

Participants also encouraged the search for PET ligands for parsing out receptor occupancy, the key feature of a structural biomarker. Other applications of PET, alone or combined with other imaging tools, also could be used to determine drug dosing: the sufficient doses to reach the target and the knowledge that can be gained about a drug's pharmacokinetics and pharmacodynamics. Another reason to broaden PET ligands is to develop molecular targets that stratify with different types of a particular disease. Most psychiatric disorders, for example, are viewed as a heterogeneous mix of disease subtypes. Because their diagnosis rests on symptoms, as opposed to biomarkers, different diagnostic groupings are unreliable, according to the prevailing wisdom among participants. Misclassification of diagnostic subgroups has serious consequences for drug development.

Linking Biomarkers to Disease Endophenotypes

For biomarker development, many participants said they prefer to focus research on endophenotypes, which are thought to be the underlying biological basis of a symptom or any other manifestation of disease. They readily lend themselves to biomarker development. In schizophrenia, for example, the endophenotype might be mismatch negativity or prepulse inhibition or a combination of the two. It might be genes or gene–gene interactions. Endophenotypes are thought to be hereditable and state independent, meaning they manifest in an affected person whether or not the symptom is active. Several workshop participants said they believed biomarkers should be linked to disease endophenotypes and be less contingent on symptom reporting. Their expectation is that specific endophenotypes might transcend descriptive disease categories and could be found in both schizophrenia and schizoaffective disorder, or in both depression and anxiety. An endophenotype might even transcend more than two disorders, encompassing as many current disorders as possible that are associated with the same genotypes. Most psychiatric disorders rely on descriptive labels and symptom criteria without regard to etiology or underlying pathophysiology. Research on functional biomarkers and treatments might progress more rapidly if endophenotypes are refined.

Public–Private Partnerships

Workshop participants indicated the potential for significant research progress on many candidate biomarkers for glutamate neurotransmission. But a concerted effort necessary to develop these candidates into valid, reliable, and widespread methods for clinical use has not yet been realized. Many participants expressed the need for collaboration and identified the success of the Alzheimer's Disease Neuroimaging Initiative (ADNI) as a potential model for such an effort. One hallmark of the ADNI effort has been the development of standardizing imaging and fluid biomarker collection across multiple sites. Many participants said that one of the major rate-limiting steps to success in biomarker translation and validation begins with uniformity and standardization of biomarkers across laboratories. For example, with respect to electrophysiology, the same equipment and ambient sound levels and light conditions would be necessary across laboratories. Development of public–private partnerships offers a potential mechanism for standardization.

One problem in developing these partnerships is the different research orientations of academia and industry. Researchers in academia tend not to be focused on drug development. Meanwhile, industry researchers have in the past viewed biomarker development as too remote from giving their company a competitive edge and commercial payoff. Simply put, glutamate-related biomarker development and validation has fallen between the cracks of academic, government, and industry research programs, participants asserted. Meanwhile, new candidate biomarkers can be developed only from large-scale collaborations that combine the resources, technology, and access to human subjects on the scale needed for glutamate biomarker development. Many of the research needs have been previously identified: the lack of PET ligands and biomarkers for patient stratification, among others. Lack of translation from animal models to humans has also been an obstacle. In short, there are no organized efforts to take advantage of existing candidate biomarkers or efforts to develop new ones. The lack of a single, validated glutamate biomarker precludes comparisons between a new medication and a benchmark. All of these factors slow progress—progress that could be accelerated with combined efforts.

SUMMARY

Despite the challenges, participants expressed enthusiasm that glutamate-related drugs will be developed in the near future, especially with coordinated investment in biomarker development. The goal is to advance understanding of the glutamate system and provide new opportunities for drug development. Biomarkers hold the key. They are vital to refining targets, they provide proof of mechanism and concept, and they help researchers to evaluate whether early intervention focused on a single target can prevent or stabilize disease, among other benefits.

Glutamate biomarkers may aid in the rational treatments of many diseases. This workshop explored the barriers to developing these biomarkers, and identified mechanisms by which those barriers can be overcome. They are only one of many new pieces in the rapidly changing world of scientific advancements to treat disease.

A

References

Bagni, C., and W. T. Greenough. 2005. From mRNP trafficking to spine dysmorphogenesis: The roots of Fragile X syndrome. *Nat Rev Neurosci* 6(5):376–387.

Barch, D. M., M. G. Berman, R. Engle, J. H. Jones, J. Jonides, A. MacDonald III, D. E. Nee, T. S. Redick, and S. R. Sponheim. 2009. CNTRICS final task selection: Working memory. *Schizophr Bull* 35(1):136–152.

Barco, A., C. H. Bailey, and E. R. Kandel. 2006. Common molecular mechanisms in explicit and implicit memory. *J Neurochem* 97(6):1520–1533.

Bechtholt-Gompf, A. J., H. V. Walther, M. A. Adams, W. A. Carlezon, D. Öngür, and B. M. Cohen. 2010. Blockade of astrocytic glutamate uptake in rats induces signs of anhedonia and impaired spatial memory. *Neuropsychopharmacology* 35:2049–2059.

Brennan, B. P., J. I. Hudson, J. E. Jensen, J. McCarthy, J. L. Roberts, A. P. Prescot, B. M. Cohen, H. G. Pope, Jr., P. F. Renshaw, and D. Öngür. 2010. Rapid enhancement of glutamatergic neurotransmission in bipolar depression following treatment with riluzole. *Neuropsychopharmacology* 35(3):834–846.

Brown, V., P. Jin, S. Ceman, J. C. Darnell, W. T. O'Donnell, S. A. Tenenbaum, X. Jin, Y. Feng, K. D. Wilkinson, J. D. Keene, R. B. Darnell, and S. T. Warren. 2001. Microarray identification of FMRP-associated brain mRNAs and altered mRNA translational profiles in fragile X syndrome. *Cell* 07(4):477–487.

Buchanan, R., M. Davis, D. Goff, M. F. Green, R. S. Keefe, A. C. Leon, K. H. Nuechterlein, T. Laughren, R. Levin, E. Stover, W. Fenton, and S. R. Marder. 2005. A summary of the FDA–NIMH–MATRICS workshop on clinical trial design for neurocognitive drugs for schizophrenia. *Schizophr Bull* 31(1):5–19.

Butler, P. D., V. Zemon, I. Schechter, A. M. Saperstein, M. J. Hoptman, K. O. Lim, N. Revheim, G. Silipo, and D. C. Javitt. 2005. Early-stage visual processing and cortical amplification deficits in schizophrenia. *Arch Gen Psychiatry* 62(5):495–504.

Cairns, B. E. 2010. Pathophysiology of TMD pain—basic mechanisms and their implications for pharmacotherapy. *J Oral Rehabil* 37(6):391–410.

Castrillon, E. E., B. E. Cairns, M. Ernberg, K. Wang, B. J. Sessle, L. Arendt-Nielsen, and P. Svensson. 2008. Effect of peripheral NMDA receptor blockade with ketamine on chronic myofascial pain in temporomandibular disorder patients: A randomized, double-blinded, placebo-controlled trial. *J Orofac Pain* 22(2):122–130.

Castrillon, E. E., M. Ernberg, B. E. Cairns, K. Wang, B. J. Sessle, L. Arendt-Nielsen, and P. Svensson. 2010. Interstitial glutamate concentration is elevated in the masseter muscle of myofascial temporomandibular disorder patients. *J Orofac Pain* 24(4):350–360.

Choi, D. W. 1994. Glutamate receptors and the induction of excitotoxic neuronal death. *Prog Brain Res* 100:47–51.

Collingridge, G. L., J. T. Isaac, and Y. T. Wang. 2004. Receptor trafficking and synaptic plasticity. *Nat Rev Neurosci* 5(12):952–962.

Cortese, B. M., and K. Luan Phan. 2005. The role of glutamate in anxiety and related disorders. *CNS Spectrums* 10(10):820–830.

Doble, A. 1999. The role of excitotoxicity in neurodegenerative disease: Implications for therapy. *Pharmacol Ther* 81(3):163–221.

Dölen, G., and M. F. Bear. 2008. Role for metabotropic glutamate receptor 5 (mGluR5) in the pathogenesis of fragile X syndrome. *J Physiol* 586(6):1503-1508.

Eulenburg, V., and J. Gomeza. 2010. Neurotransmitter transporters expressed in glial cells as regulators of synapse function. *Brain Res Rev* 63(1–2):103–112.

FDA (Food and Drug Administration). 2004. *Critical path update.* http://www.fda.gov/ScienceResearch/SpecialTopics/CriticalPathInitiative/CriticalPathOpportunitiesReports/ucm077262.htm#execsummary (accessed November 9, 2010).

FDA. 2010. *Guidance for industry: Qualification process for drug development tools.* Rockville, MD: FDA, Center for Drug Evaluation and Research. http://www.fda.gov/downloads/Drugs/GuidanceComplianceRegulatoryInformation/Guidances/UCM230597.pdf (accessed March 24, 2011).

Foran, E., and D. Trotti. 2009. Glutamate transporters and the excitotoxic path to motor neuron degeneration in amyotrophic lateral sclerosis. *Antioxid Redox Signal* 11(7):1587–1602.

Friedman, H. R., and P. S. Goldman-Rakic. 1994. Coactivation of prefrontal cortex and inferior parietal cortex in working memory tasks revealed by 2DG functional mapping in the rhesus monkey. *J Neuroscience* 14(5 Pt 1):2775–2788.

Garrido, M. I., J. M. Kilner, K. E. Stephan, and K. J. Friston. 2009. The mismatch negativity: A review of underlying mechanisms. *Clin Neurophysiol* 120(3):453–463.

Gazerani, P., S. Au, X. Dong, U. Kumar, L. Arendt-Nielsen, and B. E. Cairns. 2010. Botulinum neurotoxin type A (BoNTA) decreases the mechanical sensitivity of nociceptors and inhibits neurogenic vasodilation in a craniofacial muscle targeted for migraine prophylaxis. *Pain* 151(3):606–616.

Geyer, M. A., K. L. McIlwain, and R. Paylor. 2002. Mouse genetic models for prepulse inhibition: An early review. *Mol Psychiatry* 7(10):1039–1053.

Goldman-Rakic, P. S. 1994. Working memory dysfunction in schizophrenia. *J Neuropsychiatry Clin Neuroscience* 6(4):348–357.

Goodsaid, F., and F. Frueh. 2007. Biomarker qualification pilot process at the U.S. Food and Drug Administration. *AAPS J* 9(1):E105-E108.

Greenwood, T. A., D. L. Braff, G. A. Light, K. S. Cadenhead, M. E. Calkins, D. J. Dobie, R. Freedman, M. F. Green, R. E. Gur, R. C. Gur, J. Mintz, K. H. Nuechterlein, A. Olincy, A. D. Radant, L. J. Seidman, L. J. Siever, J. M. Silverman, W. S. Stone, N. R. Swerdlow, D. W. Tsuang, M. T. Tsuang, B. I. Turetsky, and N. J. Schork. 2007. Initial heritability analyses of endophenotypic measures for schizophrenia: The consortium on the genetics of schizophrenia. *Arch Gen Psychiatry* 64(11):1242–1250.

Grewer, C., A. Gameiro, Z. Zhang, Z. Tao, S. Braams, and T. Rauen. 2008. Glutamate forward and reverse transport: From molecular mechanism to transporter-mediated release after ischemia. *IUBMB Life* 60(9):609–619.

Hall, M. H., K. Schulze, E. Bramon, R. M. Murray, P. Sham, and F. Rijsdijk. 2006a. Genetic overlap between P300, P50, and duration mismatch negativity. *Am J Med Genet B Neuropsychiatr Genet* 141B(4):336–343.

Hall, M. H., K. Schulze, F. Rijsdijk, M. Picchioni, U. Ettinger, E. Bramon, R. Freedman, R. M. Murray, and P. Sham. 2006b. Heritability and reliability of P300, P50 and duration mismatch negativity. *Behav Genet* 36(6):845–857.

IOM (Institute of Medicine). 2008. *Neuroscience biomarkers and biosignatures: Converging technologies, emerging partnerships.* Washington, DC: The National Academies Press.

Jacobson, M. S., J. C. Hung, T. L. Mays, and B. P. Mullan. 2002. The planning and design of a new PET radiochemistry facility. *Mol Imaging Biol* 4(2):119-127.

Jahshan, C., K. S. Cadenhead, A. J. Rissling, K. Kirihara, D. L. Braff, and G. A. Light. In press. Automatic sensory information-processing abnormalities across the illness course of schizophrenia. *Psychological Medicine.*

Javitt, D. C., M. Steinschneider, C. E. Schroeder, and J. C. Arezzo. 1996. Role of cortical N-methyl-D-aspartate receptors in auditory sensory memory and mismatch negativity generation: Implications for schizophrenia. *Pro Natl Acad Sci USA* 93(21):11962–11967.

Javitt, D. C., K. M. Spencer, G. K. Thaker, G. Winterer, and M. Hajós. 2008. Neurophysiological biomarkers for drug development in schizophrenia. *Nat Rev Drug Discov* 7(1):68–83.

Johannes, C. B., T. K. Le, X. Zhou, J. A. Johnston, and R. H. Dworkin. 2010. The prevalence of chronic pain in United States adults: Results of an Internet-based survey. *J Pain* 11(11):1230–1239.

Kanai, Y., M. Stelzner, S. Nussberger, S. Khawaja, S. C. Hebert, C. P. Smith, and M. A. Hediger. 1994. The neuronal and epithelial human high affinity glutamate transporter: Insights into structure and mechanism of transport. *J Biol Chem* 269(32):20599–20606.

Kelleher, R. J., III, and M. F. Bear. 2008. The autistic neuron: Troubled translation? *Cell* 135(3):401–406.

Kew, J. N., and J. A. Kemp. 2005. Ionotropic and metabotropic glutamate receptor structure and pharmacology. *Psychopharmacology (Berl)* 179(1):4–29.

Krishnan, G. P., J. L. Vohs, W. P. Hetrick, C. A. Carroll, A. Shekhar, M. A. Bockbrader, and B. F. O'Donnell. 2005. Steady state visual evoked potential abnormalities in schizophrenia. *Clin Neurophysiol* 116(3):614–624.

Kurylo, D. D., and Y. Gazes. 2008. Effects of ketamine on perceptual grouping in rats. *Physiol Behav* 95(1–2):152–156.

Lakatos, P., G. Karmos, A. D. Mehta, I. Ulbert, and C. E. Schroeder. 2008. Entrainment of neuronal oscillations as a mechanism of attentional selection. *Science* 320(5872):110–113.

Lee, M. C., K. K. Ting, S. Adams, B. J. Brew, R. Chung, and G. J. Guillemin. 2010. Characterisation of the expression of NMDA receptors in human astrocytes. *PLoS One* 5(11):e14123.

Lesage, A., and T. Steckler. 2010. Metabotropic glutamate mGlu1 receptor stimulation and blockade: Therapeutic opportunities in psychiatric illness. *Eur J Pharmacol* 639(1–3):2–16.

Lewis, D. A., L. A. Glantz, J. N. Pierri, and R. A. Sweet. 2003. Altered cortical glutamate neurotransmission in schizophrenia: Evidence from morphological studies of pyramidal neurons. *Ann N Y Acad Sci* 1003:102–112.

Light, G. A., L. E. Williams, F. Minow, J. Sprock, A. Rissling, R. Sharp, N. R. Swerdlow, and D. L. Braff. 2010. Electroencephalography (EEG) and event-related potentials (ERPs) with human participants. *Curr Protoc Neurosci* 6(6):1–24.

Limbrick, D. D., Jr., S. Sombati, and R. J. DeLorenzo. 2003. Calcium influx constitutes the ionic basis for the maintenance of glutamate-induced extended neuronal depolarization associated with hippocampal neuronal death. *Cell Calcium* 33(2):69–81.

Malarkey, E. B., and V. Parpura. 2008. Mechanisms of glutamate release from astrocytes. *Neurochem Int* 52(1–2):142–154.

Narrow, W. E., D. S. Rae, L. N. Robins, and D. A. Regier. 2002. Revised prevalence estimates of mental disorders in the United States: Using a clinical significance criterion to reconcile 2 surveys' estimates. *Arch Gen Psychiatry* 59(2):115–123.

Niswender, C. M., and P. J. Conn. 2010. Metabotropic glutamate receptors: Physiology, pharmacology, and disease. *Ann Rev Pharmacol Toxicol* 50:295–322.

Olive, M. F. 2009. Metabotropic glutamate receptor ligands as potential therapeutics for addiction. *Curr Drug Abuse Rev* 2(1):83–98.

Pacher, P., S. Batkai, and G. Funos. 2006. The endocannabinoid system as an emerging target of pharmacotherapy. *Pharmacol Rev* 58(3):389–462.

Paulus, M. P., J. S. Feinstein, G. Castillo, A. N. Simmons, and M. B. Stein. 2005. Dose-dependent decrease of activation in bilateral amygdala and insula by lorazepam during emotion processing. *Arch Gen Psychiatry* 62:282–288.

Pike, V. W. 2009. PET radiotracers: Crossing the blood–brain barrier and surviving metabolism. *Trends in Pharmacological Sciences* 30(8):431–440.

Pleasure, D. 2008. Diagnostic and pathogenic significance of glutamate receptor autoantibodies. *Arch Neurol* 65(5):589–592.

Rothman, D. L., K. L. Behar, F. Hyder, and R. G. Shulman. 2003. In vivo NMR studies of the glutamate neurotransmitter flux and neuroenergetics: Implications for brain function. *Annu Rev Physiol* 65:401–427.

Rothstein, J. D., L. J. Martin, and R. W. Kuncl. 1992. Decreased glutamate transport by the brain and spinal cord in amyotrophic lateral sclerosis. *N Engl J Med* 326(22):1464–1468.

Salvadore, G., B. R. Cornwell, F. Sambataro, D. Latov, V. Colon-Rosario, F. Carver, T. Holyroyd, N. DiazGranados, R. Machado-Vieira, C. Grillon, W. C. Drevets, and C. A. Zarate, Jr. 2010. Anterior cingulate desynchronization and functional connectivity with the amygdale during a working memory task predict rapid antidepressant response to ketamine. *Neuropsychopharmacology* 35(7):1415–1422.

Sanacora, G., C. A. Zarate, J. H. Krystal, and H. K. Manji. 2008. Targeting the glutamatergic system to develop novel, improved therapeutics for mood disorders. *Nat Rev Drug Discov* 7(5):426–437.

Swanson, C. J., M. Bures, M. P. Johnson, A. M. Linden, J. A. Monn, and D. D. Schoepp. 2005. Metabotropic glutamate receptors as novel targets for anxiety and stress disorders. *Nat Rev Drug Discov* 4(2):131–144.

Tanaka, K. 2000. Functions of glutamate transporters in the brain. *Neuroscience Res.* 37(1):15–19.

Terry, G. E., J. Hirvonen, J. S. Liow, S. S. Zoghbi, R. Gladding, J. T. Tauscher, J. M. Schaus, L. Phebus, C. C. Felder, C. L. Morse, S. R. Donohue, V. W. Pike, C. Halldin, and R. B. Innis. 2010. Imaging and quantitation of cannabinoid CB_1 receptors in human and monkey brains using (18)F-labeled inverse agonist radioligands. *J Nucl Med* (1):112–120.

Tononi, G., and C. Cirelli. 2006. Sleep function and synaptic homeostasis. *Sleep Med Rev* 10(1):49–62.

Umbricht, D., and S. Krljes. 2005. Mismatch negativity in schizophrenia: A meta-analysis. *Schizophr Res* 76(1):1–23.

Umbricht, D., L. Schmid, R. Koller, F. X. Vollenweider, D. Hell, and D. C. Javitt. 2000. Ketamine-induced deficits in auditory and visual context-dependent processing in healthy volunteers: Implications for models of cognitive deficits in schizophrenia. *Archives Gen Psychiatry* 57(12):1139–1147.

Verrall, L., P. W. Burnet, J. F. Betts, and P. J. Harrison. 2010. The neurobiology of D-amino acid oxidase and its involvement in schizophrenia. *Mol Psychiatry* 15(2):122–137.

Wang, X. P., and H. L. Ding. 2008. Alzheimer's disease: Epidemiology, genetics, and beyond. *Neurosci Bull* 24(2):105–109.

Watson, T. D., I. L. Petrakis, J. Edgecombe, A. Perrino, J. H. Krystal, and D. H. Mathalon. 2009. Modulation of the cortical processing of novel and target stimuli by drugs affecting glutamate and GABA neurotransmission. *Int J Neuropsychopharmacol* 12(3):357–370.

WHO (World Health Organization). 2001. *The World Health Report 2001—Mental Health: New Understanding, New Hope.* Geneva, World Health Organization. http://www.who.int/whr/ (accessed March 24, 2011).

Writer, B. W., and J. E. Schillerstrom. 2009. Psychopharmacological treatment for cognitive impairment in survivors of traumatic brain injury: A critical review. *J Neuropsychiatry Clin Neurosci* 21(4):362–370.

Xiong, Z. G., G. Pignataro, M. Li, S. Y. Chang, and R. P. Simon. 2008. Acid-sensing ion channels (ASICs) as pharmacological targets for neurodegenerative diseases. *Curr Opin Pharmacol* 8(1):25–32.

Yüksel, C., and D. Öngür. 2010. Magnetic resonance spectroscopy studies of glutamate-related abnormalities in mood disorders. *Biol Psychiatry* 68:785–794.

Zarate, C. A., Jr., J. B. Singh, P. J. Carlson, N. E. Brutsche, R. Ameli, D. A. Luckenbaugh, D. S. Charney, and H. K. Manji. 2006. A randomized trial of an N-methyl-D-aspartate antagonist in treatment-resistant major depression. *Arch Gen Psychiatry* 63(8):856–864.

Zarate, C., Jr., R. Machado-Vieira, I. Henter, L. Ibrahim, N. Diazgranados, and G. Salvadore. 2010. Glutamatergic modulators: The future of treating mood disorders? *Harv Rev Psychiatry* 18(5):293–303.

B

Registered Attendees

Neha Agarwal
University of California–Berkeley

Chiiko Asanuma
National Institute of Mental
Health

Helene Benveniste
University Medical Center at Stony
Brook

Lizbet Boroughs
American Psychiatric Association

Linda Brady
National Institute of Mental
Health

Robbin Brodbeck
Lundbeck Research USA

Neil Buckholtz
National Institute on Aging

Lisheng Cai
National Institute of Mental
Health

Marc Cantillon
Coalition Against Major Disease,
Critical Path

Juan Chavez
Regeneron Pharmaceuticals

William Cho
Merck and Co., Inc.

Alice Luo Clayton
Health Journalist

Vlad Coric
Bristol-Myers Squibb

Alan Cross
AstraZeneca Pharmaceuticals

Jasbeer Dhawan
Brookhaven National Laboratory

Dario Doller
Lundbeck Research USA

Jamie Driscoll
National Institute of Mental
    Health

Melissa Dupree
Fabiani & Company

Katherine Falk
Mt. Sinai School of Medicine

Gwenaelee Fillon
Hoffman-La Roche, Inc.

Jarlath French-Mullen
GeneLogic/Ocimum Biosolutions

Robert Gibbons
University of Illinois–Chicago

Sarah Grimwood
Pfizer Inc.

Michele Hindi-Alexander
National Institutes of Health

Garry Honey
Pfizer Inc.

Thomas Insel
National Institute of Mental
    Health

Carrie Jones
Vanderbilt University

John Kehne
Translational Neuropharmacology
    Consulting, LLC

Frederick Knox
United States Navy (Retired)

Jaya Kolli
Ross University School of Medicine

David Lee
Foundation for NIH

Serguei Liachenko
National Center for Toxicology
    Research

Johan Luthman
Merck and Co., Inc.

Ronald Marcus
Bristol-Myers Squibb

Michael Marino
Merck and Co., Inc.

Shawnmarie Mayrand-Ching
National Institutes of Health

Diana Morales
National Institutes of Health

Laurie Nadler
National Institute of Mental
    Health

Richard Nakamura
National Institute of Mental
    Health

Lindsay Pascal
AAAS

Victor Pike
National Institute of Mental
    Health

Jorge Quiroz
Hoffman-La Roche, Inc.

Jean Rifkin
Diabetes Seminars

Erica Rosemond
National Institute of Mental
  Health

Deborah Runkle
AAAS

Judy Siuciak
Foundation at NIH

Renuka Sriram
Pfizer Inc.

Thomas Steckler
Johnson & Johnson

Cyrille Sur
Merck and Co., Inc.

Cristina Tan Hehir
GE Global Research

Pam Tannenbaum
Merck and Co., Inc.

Christian Thomsen
Lundbeck Research USA

Dan van Kammen
CHDI Foundation

Philip Wang
National Institute of Mental
  Health

Lois Winsky
National Institute of Mental
  Health

Frank Yocca
AstraZeneca Pharmaceuticals

Huailing Zhong
Lundbeck Research USA

C

Agenda

June 21 and 22, 2010

Institute of Medicine
500 Fifth Street, NW
Keck Building, Room 100
Washington, DC 20001

Background: Dysfunction of glutamatergic neurotransmission has been implicated in many nervous system disorders, including neuropsychiatric disorders such as anxiety, schizophrenia and major depressive disorder, and neurodegenerative diseases such as Alzheimer's disease and amyotrophic lateral sclerosis. Biomarkers that specifically measure glutamate signaling and related circuitry are a promising means to accelerate drug development for disorders associated with glutamatergic dysfunction by providing quantitative measures for diagnosis, patient stratification, and assessment of drug efficacy. The goal of the workshop is to present promising current and emerging technologies with potential as reliable glutamate biomarkers, and to outline strategies to accelerate development, validation, and implementation of these biomarkers as powerful tools to advance drug development for nervous system disorders associated with glutamatergic dysfunction.

Meeting Objectives:

- Briefly outline the need for glutamate-related biomarkers both for understanding the causes of neuropsychiatric disorders and neurodegenerative diseases associated with glutamatergic dysfunction, and for accelerating drug development for these disorders.

- Discuss the most promising current and emerging technologies and analytical methods for assessing glutamatergic neurotransmission, and identify the research gaps for their development into biomarkers.
- Outline approaches for biomarker validation in preclinical and clinical studies, including relevant animal models and translational challenges.
- Discuss the implementation and regulatory barriers to incorporating glutamatergic biomarkers into drug development for neuropsychiatric disorders and neurodegenerative diseases and approaches to overcome them.
- Identify the next steps to establish principles and procedures to accelerate biomarker development, validation, and implementation in clinical trials, including frameworks for partnerships and collaboration.

June 21: Keck Center Room 100

1:00 p.m.   Welcome, Introductions, and Workshop Objectives

CHI-MING LEE, *Workshop Cochair*
Executive Director of Translational Science
AstraZeneca Pharmaceuticals

DANIEL JAVITT, *Workshop Cochair*
Program Director
Cognitive Neuroscience and Schizophrenia
Nathan Kline Institute

1:10 p.m.   Overview of the Glutamatergic System in the CNS: How Does a Single Neurotransmitter Result in So Many Diverse Functions?

DARRYLE SCHOEPP
Senior Vice President and Franchise Head
Neuroscience
Merck

SESSION I: GLUTAMATE DYSFUNCTION IN NERVOUS SYSTEM DISORDERS AND STATEMENT OF NEED

Session Objective:
- Outline the current state of knowledge and the therapeutic gaps in several nervous system disorders associated with glutamatergic dysfunction, focusing on neuropsychiatric disorders and neurodegenerative disease. Discuss the needs and potential benefits of glutamate biomarkers for accelerating drug development for these diseases.

1:30 p.m.   Schizophrenia: Glutamate Dysfunction, Treatments, and Need for Glutamate Biomarkers

DANIEL JAVITT
Program Director
Cognitive Neuroscience and Schizophrenia
Nathan Kline Institute

1:45 p.m.   Anxiety Disorders: Glutamate Dysfunction, Treatments, and Need for Glutamate Biomarkers

JEFFREY CONN
Professor of Pharmacology
Director, Vanderbilt Program in Drug Discovery
Vanderbilt University

2:00 p.m.   Major Depressive Disorder (MDD): Glutamate Dysfunction, Treatments, and Need for Glutamate Biomarkers

CARLOS ZARATE
Chief of the Mood and Anxiety Disorders Research Unit
Associate Clinical Director of the Laboratory of Molecular Pathophysiology
NIMH, NIH

2:15 p.m.   Autism Spectrum Disorders: Glutamate Dysfunction, Treatments, and Need for Glutamate Biomarkers

MARK BEAR
Picower Professor of Neuroscience
Department of Brain and Cognitive Sciences
Massachusetts Institute of Technology
Scientific Founder, Seaside Therapeutics

2:30 p.m.   Chronic Pain: Glutamate Dysfunction, Treatments, and
Need for Glutamate Biomarkers

BRIAN CAIRNS
Associate Professor and Canada Research Chair in
Neuropharmacology
Faculty of Pharmaceutical Sciences
The University of British Columbia–Vancouver

2:45 p.m.   BREAK

3:00 p.m.   Alzheimer's Disease: Glutamate Dysfunction, Treatments,
and Need for Glutamate Biomarkers

LENNART MUCKE
Professor of Neurology
Director and Senior Investigator
Gladstone Institute of Neurological Disease
University of California–San Francisco

3:15 p.m.   Drug Addiction: Glutamate Dysfunction, Treatments, and
Need for Glutamate Biomarkers

PETER KALIVAS
Co-Chair, Department of Neurosciences
Medical University of South Carolina

3:30 p.m.   Stroke and Ischemia: Glutamate Dysfunction, Treatments,
and Need for Glutamate Biomarkers

DENNIS CHOI
Vice-President for Academic Health Affairs
Woodruff Health Sciences Center
Executive Director of the Strategic Neurosciences
Initiative
Emory University

3:45 p.m.   Amyotrophic Lateral Sclerosis (ALS): Glutamate
             Dysfunction, Treatments, and Need for Glutamate
             Biomarkers

             JEFFREY ROTHSTEIN
             Professor of Neurology
             Johns Hopkins University
             Director of the Robert Packard Foundation for ALS
             Research

4:00 p.m.   Biomarkers and Biomarker Technologies for Glutamatergic
             Therapeutics: Progress, Opportunities, and Challenges

             KALPANA MERCHANT
             Chief Scientific Officer
             Translational Science
             Eli Lilly and Company

4:20 p.m.   Discussion with Panelists and Attendees

             CHI-MING LEE, *Workshop Cochair*
             Executive Director of Translational Science
             AstraZeneca Pharmaceuticals

             DANIEL JAVITT, *Workshop Cochair*
             Program Director
             Cognitive Neuroscience and Schizophrenia
             Nathan Kline Institute

5:00 p.m.   ADJOURN

June 22: Keck Center Room 100

SESSION II: CURRENT AND EMERGING TECHNOLOGIES TO ASSESS GLUTAMATERGIC FUNCTION

Session Objectives:
- Discuss current and emerging technologies and associated analytical methods for assessing glutamatergic function, their specificity for the glutamate system, their potential for translation into clinical biomarkers, and their validation status.

- Discuss biomarkers that act on specific glutamate receptor subtypes and their relative potential as robust biomarkers.
- Present biomarkers that assess three broad aspects of glutamatergic function: function of the glutamate system (e.g., imaging), mutations in genes involved in glutamate system function, and gene expression (mRNA and protein) in the glutamate signaling pathway. Discuss their relative potential as robust biomarkers.
- Identify opportunities for increasing biomarker sensitivity using combinatorial approaches (e.g., combining proteomics and imaging).
- Discuss approaches to standardize and optimize these approaches as glutamate biomarkers, including sample collection, data acquisition, and analysis.
- Identify the scientific gaps and barriers to implementing these approaches as biomarkers for clinical research, and strategies for overcoming them.

9:00 a.m.   Welcome and Review of Day One

                CHI-MING LEE, *Workshop Cochair*
                Executive Director of Translational Science
                AstraZeneca Pharmaceuticals

                DANIEL JAVITT, *Workshop Cochair*
                Program Director
                Cognitive Neuroscience and Schizophrenia
                Nathan Kline Institute

9:10 a.m.   Panel Discussion: Sensory-Based Biomarkers

                DANIEL UMBRICHT, *Panel Chair*
                Lead, Neuroscience Translational Medicine
                Hoffmann-La Roche

                Auditory Sensory Responses—Mismatch Negativity; N1; Auditory Steady-State Responses

                GREGORY LIGHT
                Associate Professor of Psychiatry
                University of California–San Diego

Visual Measures—Transient and Steady-State Visually Evoked Potentials (ssVEP); SPEM (Smooth Pursuit Eye Movements); Visual P1

BRIAN O'DONNELL
Professor of Psychology
Indiana University

Gating Measures—Auditory P50 Response, Prepulse Inhibition

BRUCE TURETSKY
Associate Professor of Psychiatry
Associate Director, Neuropsychiatry Division
Director, Neurophysiology and Brain Imaging Laboratory
University of Pennsylvania

10:00 a.m.   Discussion with Panelists and Attendees
• What potential targets have the potential for treatment and what is their status?
• What biomarkers currently could be used to evaluate medications that target the glutamatergic system?
• What biomarkers have the potential to be used clinically to guide glutamatergic medications?

DANIEL UMBRICHT, *Panel Chair*
Lead, Neuroscience Translational Medicine
Hoffmann-La Roche

10:30 a.m.   BREAK

10:45 a.m.   Panel Discussion: Cognition-Based Biomarkers

MIHALY HAJOS, *Panel Chair*
Neuroscience Department, CNS Discovery
Pfizer Global Research and Development

Error-Related Negativity (ERN)

CINDY YEE-BRADBURY
Associate Professor
Department of Psychology
University of California–Los Angeles

P300 (Auditory; Visual)

DANIEL MATHALON
Associate Adjunct Professor
Department of Psychiatry
Codirector of the Brain Imaging and EEG Laboratory
University of California–San Francisco

Working Memory—Event-Related Potential (ERP);
Functional MRI (fMRI); Gamma-Evoked Oscillations

ANGUS MACDONALD
Associate Professor
Department of Psychology
University of Minnesota

Hippocampal Function and Long-Term Potentiation (LTP)

JOHN LISMAN
Professor of Biology
Brandeis University

11:45 a.m.   Discussion with Panelists and Attendees
- What potential targets have the potential for treatment and what is their status?
- What biomarkers currently could be used to evaluate medications that target the glutamatergic system?
- What biomarkers have the potential to be used clinically to guide glutamatergic medications?

MIHALY HAJOS, *Panel Chair*
Neuroscience Department
CNS Discovery
Pfizer Global Research and Development

12:15 p.m.   LUNCH

12:45 p.m.   Panel Discussion: Genetic, Biochemical, and Metabolic-Based Biomarkers

NORA VOLKOW, *Panel Chair*
Director
National Institute on Drug Abuse

Positron Emission Tomography (PET) Approaches to the Glutamatergic System (Pre-and Post-Synaptic Receptors, Transporters)

ROBERT INNIS
Molecular Imaging Branch
PET Neuroimaging Sciences Section
NIMH, NIH

An Integrated Imaging and Animal Models Approach to Identifying Schizophrenia Biomarkers

STEPHEN RAYPORT
Associate Professor of Clinical Neuroscience
Department of Psychiatry
Columbia University Medical Center

Proton Magnetic Resonance Spectroscopy (MRS)

ROBERT MATHER
Principal Scientist
Pfizer Pharmaceuticals

Pharmacogenomic and Epigenomic Markers
(Disease-Specific)

ANNE WEST
Assistant Professor
Department of Neurobiology
Duke University

Metabolomics in the Study of Neuropsychiatric Diseases

RIMA KADDURAH-DAOUK
Associate Professor
Department of Psychiatry
Duke University

2:25 p.m.   Discussion with Panelists and Attendees
- What potential targets have the potential for treatment and what is their status?
- What biomarkers currently could be used to evaluate medications that target the glutamatergic system?
- What biomarkers have the potential to be used clinically to guide glutamatergic medications?

NORA VOLKOW, *Panel Chair*
Director
National Institute on Drug Abuse

3:00 p.m.   BREAK

SESSION III: GLUTAMATE BIOMARKER VALIDATION: ANIMAL MODELS AND CLINICAL TRIAL DESIGN

Session Objectives:
- Discuss the key factors that will affect successful translation of rodent and primate biomarkers.
- Examine the advantages and limitations of rodent models in biomarker development and validation, and examine approaches to leverage rodent models for biomarker development and for biomarker-based preclinical trials.
- Explore opportunities that were discussed that could facilitate combinatorial approaches (e.g., combining proteomics and imaging).
- Identify the key principles that should be considered with biomarker validation in primate models, including most relevant technologies for testing in primates.
- Identify and discuss optimal principles of clinical trial design for biomarker validation in humans.

3:15 p.m.   Session Objectives and Introduction

WILLIAM POTTER, *Panel Chair*
Cochair Emeritus,
Neuroscience Steering Committee, FNIH Biomarkers
    Consortium
Former Vice President, Translational Neuroscience,
Merck & Co., Inc.

3:20 p.m.   Biomarkers in Rodent Models: Challenges in Preclinical to Clinical Translation of Glutamate Biomarkers and Strategies to Overcome Them

    MARK GEYER
    Professor of Psychiatry
    University of California–San Diego

3:35 p.m.   Biomarker Validation in Primates: Relevant Technologies and Translational Considerations

    CHARLES SCHROEDER
    Laboratory for Cognitive Neuroscience and Neuroimaging
    Nathan Kline Institute

3:50 p.m.   Biomarker Validation in Humans: Roles of Healthy Volunteer and Patient Studies; Approaches to Cross-Site Standardization

    GUNVANT THAKER
    Professor
    Department of Psychiatry
    University of Maryland School of Medicine

4:05 p.m.   Discussion with Speakers and Attendees

    WILLIAM POTTER, *Panel Chair*
    Cochair Emeritus,
    Neuroscience Steering Committee, FNIH Biomarkers
        Consortium
    Former Vice President, Translational Neuroscience,
    Merck & Co., Inc.

SESSION IV: NEXT STEPS FOR ACCELERATING GLATAMATE BIOMARKER DEVELOPMENT AND QUALIFICATION

Session Objectives:
- Discuss the current barriers to biomarker development in academia and industry.
- Discuss the opportunities for partnerships to advance glutamate biomarker development (e.g., public–private partnerships in the precompetitive space) within existing or novel frameworks.

- Discuss strategies for pooling of resources and data, including failed clinical trial data.
- Identify a path forward to establish principles and procedures leading to the qualification of glutamate biomarkers in order to address unmet medical needs in drug development for diseases associated with glutamatergic dysfunction.

4:35 p.m.   *Moderated Discussion*:

STEVEN PAUL, *Panel Chair*
Executive Vice President, Science (Former)
Eli Lilly

Panel Discussion:

DENNIS CHOI
Vice President for Academic Health Affairs
Woodruff Health Sciences Center
Executive Director of the Strategic Neurosciences
    Initiative
Emory University

MARK BEAR
Picower Professor of Neuroscience
Department of Brain and Cognitive Sciences
Massachusetts Institute of Technology
Scientific Founder, Seaside Therapeutics

DANIEL UMBRICHT
Lead, Neuroscience Translational Medicine
Hoffmann-La Roche

JOHN DUNLOP
Senior Vice President and Chief Scientific Officer
Neuroscience Research Unit
Pfizer

STEVIN ZORN
Executive Vice President, R&D
Lundbeck

DARRYLE SCHOEPP
Senior Vice President and Franchise Head
Neuroscience
Merck

Discussion Questions:
- What are the most promising technologies and analytical methods for assessing glutamatergic function with potential for development and translation into robust clinical biomarkers?
- What are the implementation barriers to incorporating glutamatergic biomarkers into drug development for neuropsychiatric disorders and neurodegenerative diseases and approaches to overcome them?
- What partnerships are needed to accelerate glutamate biomarker research and development for translational medicine?
- What are the next steps toward establishing research principles and guidelines for qualification of glutamatergic biomarkers?

5:15 p.m.   Discussion with Speakers and Attendees

5:30 p.m.   ADJOURN